GLOBAL LAWS ON USE OF AI IN AUTONOMOUS VEHICLES

VETERAN WG CDR (DR) SK PRAKASH

INDIA • SINGAPORE • MALAYSIA

CONTENTS

ACKNOWLEDGEMENT

The contents of this book are part of my PhD thesis which was a dream come true. I could achieve this dream with the help and support of many people. I would like to thank and acknowledge all those who have been involved in the making of my doctoral research.

First and foremost, I would like to thank my PhD supervisor and guide Dr. V Shyam Kishore Professor and Associate Dean of Alliance School of Law, Alliance University, Bangalore for the valuable guidance throughout my research and for supporting me whenever I needed help in all aspects. I would also thank my Dean, Dr. Kiran Gardner for supporting me from time to time during progress report meetings and by guiding me to various reading materials on my research topic.

I would also like to thank the various Doctoral Progress Committee members from Alliance School of Law, Alliance University for reviewing my progress continuously and giving me their valuable feedback on my research topic.

Thanks to my PhD. Batch co-scholars who all could be considered as my critical reviewer(s), and have been very kind enough to support and motivate me throughout my research work.

I would also like to thank my father **Late Mr. Krishna Swamy Moorthy** and my mother **Mrs. Rajeswari Moorthy** for motivating me in every aspect of life. Finally, my heartfelt appreciation to my loving wife **Mrs. Sita Lakshmy Prakash,** my sweet daughter **Shweta Prakash** and my loving son **Aditya Prakash.** Without their continuous support and motivation, I would never have completed my research.

Veteran Wg Cdr (Dr) SK Prakash

ABSTRACT

Artificial Intelligence or AI is classified into three types. The first type is called narrow AI. This is also known as weak/applied AI. Narrow AI normally is linked with functions like data processing. The second type of AI is artificial general intelligence known in short as AGI or "strong" AI. AGI can replicate human thought. The last type of AI is that, which is even better than the best human brains in almost any field. This kind of AI is called artificial super intelligence (ASI). As of date, AIs are generally designed to solve the immediate tasks at hand. AGI development is unheard of, and some companies may be researching that in stealth mode. AI has slowly invaded our lives in many ways and they are performing tasks that in the recent past were not within the capability of a human to perform, unless that human had extensive training. It is no longer a secret that the USA has permitted road testing of self-driving autonomous vehicles (AVs) in twenty-nine states of the United States of America and the District of Columbia. The arrival of self-driving AVs is going to change the road transport landscape.

One of the core issues that comes to one's mind regarding self-driving AVs is the liability for accidents/violations caused by these self-driving AVs. Normally in any accident caused by self-driving AVs, there will be multiple persons responsible or liable e.g., the Maker of the AI, the Developer, the Quality Control team, the User, the person responsible for software updates, malware patches, etc. Some **sub-issues** that would arise when using self-driving AVs are:

(a) Vehicle Testing Regulations for both closed spaces and public roads,

(b) Autonomous Vehicle Driver License Eligibility Criteria,

(c) Amendment to Insurance Liability aspects for accidents caused by Autonomous Vehicles.

This book has carried out a deep dive into the above issues and analysed the existing motor vehicle and related laws in India to recommend changes needed in various Indian statutes and rules for allowing self-driving automated vehicles to ply in India.

This book consists of six chapters. The first chapter deals with an introduction to AI, a description of various types of AI, and an introduction to existing legal scenarios and the relevance of AI from global, national (India), and individual perspectives. The second chapter deals with the existing legal status of AVs in the USA and Major EU Nations and covers the existing laws, regulations, and policies on AVs in the USA along with important/landmark case laws. The third chapter deals with the existing legal status of AVs in Asia Pacific Japan and China (APJC), Australia and New Zealand (ANZ), and the Rest of Europe and covers the existing laws, regulations, and policies on AVs in APJC, ANZ, and the Rest of Europe along with important/landmark case laws. The fourth chapter contains an analysis of various legal provisions concerning AVs like Vehicle Platooning, Autonomous Vehicle Testing including the existing legal framework, speed limits, rules for engaging autopilot, and the requirement of permits, and insurance for AVs. In addition, the chapter also analyses provisions concerning infrastructure for plying AVs like mandatory road markings, interactive signs, and signals, driver's license for driving AVs in the car or remotely, product liability, usage of vehicle data, usage of AVs as public transports, recognition of various Society of Automotive Engineers (SAE) levels of automation, the liability regime for AVs (includes penalty limits) in the three major continents of the world United States of America (USA), Europe, and Asia.

The fifth chapter deals with changes required in existing laws and policies in India to permit Autonomous Vehicles to be tested/ plied on public roads in India. This chapter analyses the gaps in existing Indian penal laws in comparison with the rest of the world for permitting Autonomous Vehicles to be tested/plied on public roads, gaps in the existing Indian Evidence Act (presently Bharatiya Sakshya Adhiniyam) and Information Technology Act in comparison with the rest of the world for permitting Autonomous Vehicles to be tested/plied on public roads and finally also discusses the concept of Product Liability in India, its comparison with rest of the world and gaps to be plugged to permit Autonomous Vehicles to be tested/plied on public roads in India. The sixth chapter is the concluding chapter which summarises the rights and liabilities when using Artificial Intelligence in Autonomous Vehicles.

***Keywords*:** **Autonomous Vehicles, AI, ASI, AGI, AI Liability, Human, Self-driving cars, Truck Platooning, Driverless cars in India.**

LIST OF CASES

AUTONOMOUS VEHICLE ACCIDENT CASES
(Referred to in the thesis)

FATAL

1.	Uber Self-Driving Vehicle Crash in Tempe, Arizona	- 18 March 2018
2.	Williston Florida Tesla Crash	- 07 May 2016
3.	Mountain View, California Tesla Crash	- 23 March 2018
4.	Delray Beach, Florida Tesla Crash	- 01 March 2019
5.	Arendal, Norway Tesla Fatal Crash	- 29 May 2020
6.	Handan, China Tesla Crash	- 20 January 2016
7.	Kanagawa, Japan Tesla Crash	- 29 April 2018
8.	Guangdong., China Tesla Model Y Crash	- 05 November 2022

NON-FATAL

1.	Culver City, California, Tesla Model S	- 22 January 2018
2.	South Jordan, Utah, Tesla Model S	- 11 May 2018
3.	Moscow Ring Road in Moscow, Russia, Tesla Model 3	- 10 August 2019
4.	Taiwan, Tesla	- 01 June 2020
5.	Cruise – Prius Crash SFO, California	- 03 June 2022

Product Liability AND Other related Liability Cases

1. A. K. Roy & Another vs Voltas Limited – AC making noise, once replaced no more remedy, 1973 AIR 225.

2. A.S. Mittal v. State of U.P. (1989) – Mass Eye Surgery case – Was a pre-test of eye saline done? 1989 AIR 1570.

3. Airbus Industrie v. Laura Howell Linton (1994) - Court agreed to handle the product liability case even though we had no law on that in India at that time, ILR 1994 KAR 1370.

4. Atul Anant Rane & Ors. Vs. Lourdes Gas Service & Ors – Death due to wrong handling of the Gas knob – No Remedy, Complaint Nos. 34 & 288 of 1993, decided on, 04 November 1997 at, Maharashtra State Consumer Disputes Redressal Commission Mumbai.

5. EID (Parry) India Ltd. v. Baby Benjamin Tushara – Damage to commode after 30 years of use, 1992 (23) DRJ 394.

6. Escola v. Coca Cola Bottling Co. (1944) – Manufacturer responsible for Defective Product, 24 Cal.2d 453, 150 P.2d 436 (Cal. 1944).

7. G. Govindan vs New India Assurance Co. Ltd. on 8 April 1999, 1999 (2) SCR 476.

8. Greenman v. Yuba Power Products, Inc and Henningsen v. Bloomfield Motors, Inc – Manufacturer responsible for defect/damage if the product was used as per specification, 59 Cal.2d 57 (Cal. 1963) and 32 N.J. 358 (N.J. 1960).

9. Justice K.S. Puttaswamy (Retd) vs Union of India and Ors. on 24 August 2017), AIR 2017 SC 4161.

10. K. Gopalakrishnan Minor vs Sankara Narayanan and Ors. on 4 October, 1967, AIR 1968 Mad 436.

11. MacPherson v. Buick Motor Co (1916) – Wooden Wheel Collapsed – No Privity of Contract ruled, 111 N.E.1050 (1916).

12. Madineni Kondaiah & Ors. v. Yaseen Fatima & Ors. on June 22, 1985, 1 (1986) ACC 501.

13. MC Mehta & Another versus Union of India in the Bhopal gas case,1987 AIR 1086

14. M/s. Solidaire India Ltd. v/s D.A. Mohan Kumar – TV Picture tube damaged after the warranty period. No remedy but the company gave a discount for a new tube - Appeal No. 18 of 1990, decided on, 29 June 1991 at, Kerala State Consumer Disputes Redressal Commission Thiruvananthapuram.

15. National Insurance Co. Ltd. vs Deorao and Ors. on 8 July, 2002, II (2004) ACC 328.

16. National Insurance Co. Ltd. vs Faqir Chand and Ors. on 25 April, 1994, AIR 1995 J K 91.

17. Oriental Insurance Co. Ltd vs Inderjit Kaur & Ors on 8 December 1997, 1998 SCC (1) 371.

18. Sohan Lal Passi vs P. Sesh Reddy & Ors on 17 July 1996, 1996 SCC (5) 21.

19. S. Sudhakaran vs A.K. Francis and Ors. on 12 June 1996, AIR 1997 Ker 26.

20. Winterbottom v. Wright (1842) – Contract for Mail Coach Service Privity of Contract Case, (1842) 10 M&W 109).

LIST OF ABBREVIATION

A $	Australian Dollar
AAMVA	American Association of Motor Vehicle Administrators
AB	Assembly Bill
ACC	Allahabad Criminal Cases
ADAS	Advanced Driver Assistance Systems
ADS	Automated Driving System
ADU	Autonomous Driving Units
AEB	Automated Emergency Braking
AGI	Artificial General Intelligence
AI	Artificial Intelligence
AIR	All India Reporter
ALKS	Automated Lane Keeping System
ANI	Artificial Narrow Intelligence
ANZ	Australia and New Zealand
AP	Andhra Pradesh
APJC	Asia Pacific Japan China
ASEAN	Association of South East Asia Nation
ASI	Artificial Super Intelligence
AV	Autonomous Vehicles
AV START	American Vision for Safer Transportation through Advancement of Revolutionary Technologies

AVRI	Autonomous Vehicles Readiness Index
BCDA	Bases Conversion and Development Authority
BMV, USA	Bureau of Motor Vehicles, USA
C4IR	Centre for the Fourth Industrial Revolution
CAV	Centre for Connected and Autonomous Vehicles
CBU	Completely Built Up
CCTV	Closed Circuit Television
CEF	Connecting Europe Facility
CEO	Chief Executive Officer
CETRAN	Centre of Excellence for Testing and Research of Autonomous Vehicles at Nanyang Technological University
C-ITS	Cooperative Intelligence Transport Systems
CKD	Completely Knocked Down
CPA	Consumer Protection Act
DC	District of Columbia
DGT	Direccion General de Trafico
DHSMV	Department of Highway Safety and Motor Vehicles
DOT	Department of Transportation
DP	Data Protection
DVD	Digital Video Disc
DVR	Department of Vehicle Regulation
EC	European Commission
E.G.	Exempli Gratia- For Example
eKTP	Kartu Tanda Penduduk Elektronik - Indonesian Electronic Residential Card
EU	European Union
EUR	EURO
EV	Electric Vehicle
FA Act	Fatal Accidents Act 1855
FAA	Federal Aviation Administration

FAVP	Federal Automated Vehicles Policy
FCA	Fiat Chrysler Automobiles
FCC	Federal Communications Commissions
FDA	Food and Drug Administration
FDOT	Florida Department of Transportation
FEDRO	Federal Roads Office - Switzerland
GM	General Motors
GPS	Global Positioning System
GTR	Global Technical Regulations
HAL	Hindustan Aeronautics Ltd
HAV	Highly Automated Vehicles
HB	House Bill
HGV	Heavy Goods Vehicle
HMV	Heavy Motor Vehicle
IA	Indian Airlines
ICV	Intelligent Connected Vehicles
Insurance Code	Bulgarian Insurance Code 2016
IP	Intellectual Property
IPC	Indian Penal Code (presently Bharatiya Nyaya Sanhita)
ISO	International Organisation for Standardisation
IT	Information Technology
IT Act 2000	Information Technology Act 2000
IV	Infrastructure Victoria
J&K	Jammu and Kashmir
JV	Joint Venture
Ker	Kerala
KPMG	Klynveld Peat Marwick and Goerdele
KT	Korea Telecom
LB	Legislative Bill
LiDAR	Light Detection and Ranging

LMV	Light Motor Vehicle
LTA	Land Transport Authority of Singapore
M&M	Mahindra and Mahindra
MaaS	Mobility as a Service
MACT	Motor Accident Claims Tribunal
MIT	Massachusetts Institute of Technology
MLIT	Ministry of Land, Infrastructure, Transport and Tourism
MOCI	Minister of Communication and Informatics
MOT	Ministry of Transport and Road Safety
MOU	Memorandum of Understanding
MRTP	Monopolies and Restrictive Trade Practices
MSP	Model State Policy
MV Act	Motor Vehicles Act 1988 - india
NDL	National Digital Library
NHAI	National Highways Authority of India
NHTSA	National Highway Traffic Safety Administration.
NITI	National Institution for Transforming India
NLSIU	National Law School of India University
NoVA	North Virginia
NSC	National Safety Council
NTSB	National Transportation Safety Board
ODD	Operational Domain Design
OEM	Original Equipment Manufacturer
OZ	Slovenian Obligations Code - Obligacijski zakonik
POC	Proof of Concept
PRC	Peoples Republic of China
PrHG	*Produktehaftpflichtgeset - Swiss Product Liability Act*
R&D	Research and Development
RDW	Netherlands Vehicle Authority

RMB	Renminbi
Road Traffic Act	Bulgaria Road Traffic Act 1999
RPAS	Remotely Piloted Aerial Systems
RTA	Road Traffic Act
SAE	Society of Automotive Engineers
SAV	Shared Autonomous Vehicle
SB	Senate Bill
SC	Supreme Court
SELF DRIVE	Safely Ensuring Lives Future Deployment and Research in Vehicle Evolution
STA	Singapore Road Traffic Act
StVG	*Straßenverkehrsgesetz - The German Road Traffic Act*
SVG	Federal Road Traffic Act - Switzerland
TNO	Netherlands Organisation for Applied Scientific Research
TP	Third Party
TV	Television
U.P	Uttar Pradesh
UAS	Unmanned Aerial Systems
UK	United Kingdom
UNECE	United Nations Economic Commission for Europe
UNSDG	United Nations Sustainable Development Goals
US	United States
USA	United States of America
USD	United States Dollar
USDOT	United States Department of Transportation
V2V	Vehicle to Vehicle
VC	Vienna Convention
VTTI	Virginia Tech's Transportation Institute
ZPrCP	Zakon o pravilih cestnega prometa - Slovenian Road Traffic Rules

CHAPTER 1

INTRODUCTION

1.1 Introduction

Artificial Intelligence is a term that refers to the capability of computers or robots to perform tasks that a normal human would be able to perform.[1] Some of these tasks are speech recognition, language translation, decision-making, and visual perception[2]. Artificial Intelligence or AI is basically classified into three types namely narrow AI, artificial general intelligence or AGI, and artificial super intelligence or ASI[3]. Machine intelligence that is greater than human intelligence across all tasks is called ASI, whereas AGI or strong AI is the term used to describe machine intelligence that is capable of performing all tasks at a level that is comparable to that of humans[4].

As of date, AIs are generally designed to solve the tasks at hand. AGI development is unheard of, though some companies may be researching that in stealth mode. AI has slowly invaded our lives in many ways, and they are performing tasks that in the recent past were not within the capability of a

---

[1] Artificial Intelligence [online] available at: https://www.britannica.com/technology/artificial-intelligence [last visited on February 17, 2022]. Per this article, Artificial Intelligence is a term that is frequently applied to the project of developing systems endowed with the intellectual processes similar to humans.

[2] What is artificial intelligence [online] available at https://marketbusinessnews.com/financial-glossary/artificial-intelligence/ [last visited on February 17, 2022]. Per this article, Artificial Intelligence, or AI has several applications, including software simulations and robotics.

[3] Stephen De Spiegeleire, Matthijs Maas & Tim Sweijs, 2 What is Artificial Intelligence? - JSTOR, https://www.jstor.org/stable/pdf/resrep12564.7.pdf (last visited February 17, 2022).

[4] Id

human to perform, unless that human had extensive training. Some of the various tasks that an AI can perform are creating perfumes through various combinations, food menus, fashion wear, functioning as a lyricist or movie maker, fortune telling and the latest addition is beauty AI which helps judge beauty contestants.

There is quite a buzz around artificial attorneys hired by some law firms, and robotic surgery being done in some hospitals. Most smartphones are Artificial Intelligence (AI) enabled to understand the applications that we use most, and they also predict the direction where our car will move when we start it. AI is the science that helps to make machines as intelligent as humans[5]. Present-day AI, works on big data and can perform most tasks a human[6] can do. These AI's have the capability to take unforeseeable decisions.

AI in recent years is also the brain[7] that powers autonomous vehicles or AVs. AI helps self-driving AVs drive safely and avoid accidents just like any sane human driver. It is no longer a secret that the USA has permitted road testing of self-driving AVs in many states and this number is increasing year after year[8]. In 2018, fifteen states enacted eighteen bills on AVs. This was preceded by the introduction of similar bills by various states starting from the year 2012[9]. In the upcoming years, the landscape of road transportation is anticipated to shift with the introduction of self-driving AVs.

One of the core issues that comes to one's mind regarding self-driving AVs is the liability for accidents/violations caused by these self-driving AVs. Normally in any accident caused by self-driving AV, there will be multiple persons responsible or liable such as the person who created the AI, the developer, the quality control team, the user, and the person in charge of virus patches and software updates.

---

[5] Artificial Intelligence. Supra Note 1, at 1.

[6] What is artificial intelligence Supra Note 2, at 1. Per this article, humans can *learn as we go along.'* In other words, learn from experience. Machines with AI can also do that.

[7] Stephen De Spiegeleire, Matthijs Maas & Tim Sweijs, What is Artificial Intelligence? Supra Note 3, at 1. The article written for Haque centre for Strategic studies describes Artificial Narrow Intelligence (ANI or "narrow AI") as machine intelligence that equals or exceeds human intelligence for specific tasks.

[8] NCSL-Autonomous Vehicles-Self Driving Vehicles Legislations Enacted [online] available at https://www.ncsl.org/research/transportation/autonomous-vehicles-self-driving-vehicles-enacted-legislation.aspx (last visited February 17, 2022). The article mentions that each year, the number of states considering legislation related to autonomous vehicles has gradually increased.

[9] Id

Several related situations that could occur when using self-driving AVs are vehicle testing regulations for both closed space and public roads, autonomous vehicle driver license eligibility criteria, an amendment to insurance liability aspects for accidents caused by AVs, and product liability. This book has carried out a deep dive into the above issues and analysed the existing motor vehicle and related laws in India to recommend changes needed in various Indian acts and rules for allowing self-driving automated vehicles to ply in India.

1.2   Basic Concepts of Autonomous Vehicles

Autonomous driving is no longer fiction. It has moved away from being science fiction to reality in the last two-plus decades. This has been possible due to progressive developments in the areas of microprocessors and radars[10]. Also, there has been a great deal of advancement in the development of portable and light technology. This has, in turn, helped to develop ultra-light hardware to facilitate quick decision-making in real-time scenarios.

James Adams a student at Stanford was the first to develop a completely autonomous lunar rover in the year 1961[11]. James eventually invented the world's first truly self-driving vehicle which was called the "Cart", which had a lot of cameras fitted to help it follow a straight white line on the ground. Thereafter, this kind of technology was considered a niche area till Tesla in 2015 came up with Autopilot in their Model S sedan[12]. This Tesla vehicle having twelve ultrasonic sensors was designed to navigate all kinds of roads and signals with minuscule involvement of the driver[13]. This development gave hope that such a technology could reach the hands of mass-market consumers.

Akin to the basic Tesla model, the present-day self-driving autonomous vehicles or AVs use a combination of cameras and radars to collect all required information about the road and then process the same to ensure safe driving. The intelligence in the software helps the AVs to correctly detect objects and decide the best path to ensure the least resistance and avoidance of collision. To provide classification to the AVs, SAE International, or the Society of

---

[10] A brief history of Autonomous Vehicle Technology [online] available at https://www.wired.com/brandlab/2016/03/a-brief-history-of-autonomous-vehicle-technology/ (last visited February 17, 2022).

[11] Id.

[12] Tesla rolls out autopilot technology [online] available at https://www.cnbc.com/2015/10/14/tesla-rolls-out-autopilot-technology.html (last visited February 17, 2022).

[13] Id.

Automotive Engineers[14] has come up with six stages of automation, level 0, which denotes complete automation, through level 5, which denotes no automation[15]. As of date, testing is mainly on Levels 2 and 3 automations while the race towards Levels 4 and 5 is in full force.

1.3   Regulatory Gaps

A vehicle is a mechanically or electrically propelled object which is used for moving goods or transporting people, especially on land. Some examples are cars, buses, scooters, trucks, and vans. Vehicles could be personal vehicles like cars, or scooters, or commercial vehicles like trucks, or buses. Vehicles could also be classified as hazardous vehicles based on what is transported in them like fuel tankers, and chemical tankers. Across the world, various motor vehicle laws apply to vehicles that are plying on roads and the drivers of these vehicles must ensure compliance with these laws. Also, the laws are different for different classes of vehicles. One example is the categorization of driving licenses based on vehicle type. A person holding a license to drive a car cannot drive a truck unless he/she has the license to drive a car and truck.

Autonomous Vehicles are of different types as classified by SAE levels and for the ones where the presence and involvement of the driver are mandatory, the applicable motor vehicle laws would be the same as that for any conventional motor vehicle. However, it is common sense that the same law cannot be applied to self-driving AVs or driverless vehicles and therefore the existing law needs to be tweaked or amended accordingly. The gaps must be identified, and amendments made to the law.

As self-driving AVs evolve policymakers across the world, including India, will have to understand the existing laws, their limitations, and the necessity to pass new laws/ update existing laws. Most countries across the world are generally in unison and feel that updating existing laws after doing a gap analysis would be the ideal and prudent thing to do instead of reinventing the wheel completely.

---

[14] Public Interest Law Journal of New Zealand available at http://www.nzlii.org/nz/journals/NZPubIntLawJl/2020/3.html / (last visited on September 19, 2022)

[15] SAE levels of driving automation [online] available at https://www.sae.org/blog/sae-j3016-update (last visited February 17, 2022).

Regulatory gaps can generally be classified under areas of targeting, novelty, obsolescence, and uncertainty[16]. Novel regulatory gaps mean a situation, where technology requires bespoke action by the government. A noteworthy example is the Standard set by the Food and Drug Administration (FDA) to certify safety for certain high-risk medical equipment in the healthcare segment[17]. The second kind of gap is an obsolete regulatory gap. This means that the existing regulation is rendered irrelevant with the development of technology. One example is an incentive for safe drivers based on hours of safe driving[18]. This will become obsolete if self-driving AVs are introduced as owners of these AVs could boost safe driving without driving a mile.

The third type is the targeting[19] gap which can be of two types namely inclusion and under-inclusion. Over-inclusion means including a technological development under a regulation even though such inclusion has no scope to further its objective, which is just the opposite of under-inclusion[20]. This indicates that although incorporating the technology into regulations will advance the objective, it is not done[21]. The best example is how Colorado does not differentiate between drunken driving by drivers of normal cars and self-driving AVs. The last kind of gap is uncertainty[22] where it is difficult to classify new technology under the right regulation. Understanding the gaps in laws required for plying normal vehicles vis-à-vis AVs has been a core objective of this book.

1.4 Limitations

Artificial Intelligence or AI is the buzzword today and most technologies/ corporations are trying to imbibe every bit of AI possible in their domain

---

[16] Can existing laws cope with AI Revolution [online] available at https://www.brookings.edu/techstream/can-existing-laws-cope-with-the-ai-revolution/ (last visited February 19, 2022).

[17] Id. As per this article, significantly, of the regulatory gaps identified, only 12 percent represent novel challenges that compel government action through the creation or adaptation of regulation.

[18] Id. This article also states that another 20 percent of the gaps are cases in which AI has made or will make regulations obsolete.

[19] Id. The article mentions that a quarter of the gaps are problems of targeting, in which regulations are either inappropriately applied to AI or miss cases in which they should be applied.

[20] Id.

[21] Id.

[22] Id. The largest group of regulatory gaps are ones of uncertainty in which a new technology is difficult to classify, causing a lack of clarity about the application of existing regulations

to ensure fast-track business growth and shape the future of tomorrow. AI has applicability in almost all facets of our daily life be it from agriculture to manufacturing, from profiling through big data to predicting election results, and from healthcare to AVs. Each of these facets has multiple sub facets and each sub-facet itself can constitute a separate area of study.

Autonomous Vehicle is a growing area of law, and the Author has made an honest attempt to look into all available data/laws concerning AVs as of date to finally narrow down on the main focus area i.e., to specifically identify the required changes in the laws of India to permit testing and usage of AVs in India. To achieve this end, the Author has focused on identifying the necessary legal foundation for allowing AV testing and usage, comparison of existing global laws on AVs concerning testing, usage, recording of evidence and liability and lastly bringing out the gaps in the existing Indian laws which need to be fixed before permitting testing and usage of AVs on public roads in India.

1.5 Importance of studying AI in Autonomous Vehicles

It has been a great human dream for many years to create an artificial person who can act like a human in all aspects. With the progress in technology, that dream is now looking to turn into reality soon. India has always been a frontrunner in adopting the best technologies introduced in the world e.g., we all know that in the telecom area especially the mobile side, we were slow starters, but now we are ready with 5G along with advanced countries across the world. Similarly, the day is not far when the adoption of AVs shall also become a prime necessity in India. If we work on the gaps in the legal regime now, allowing AVs to ply will then be relatively easy.

Numerous AIs across the world have been involved in accidents of deadly character. In most of these accidents contributions made by the AI itself are questionable. As a result, a great amount of noise has been made especially considering that there could be crimes in the future committed by AI, where it could be difficult to attribute culpability to a human. This scenario can also occur during the use of self-driven AVs if the culpability on account of the failure of different processes involved in the running of such AVs is not properly prescribed. In the absence of proper guidance and laws governing the responsibility of AI's behaviour, stretching human responsibility over AI can be a great challenge.

If one delves into past research on AI, the research is negligible concerning criminal law, other than Hallevy's work on the criminal responsibility of AI systems[23] together with a few works by Karnow, Grimm, Smart, and Hartzog[24]. Even if the problems AI poses are even more problematic in criminal law, the majority of scholars have been focusing on civil responsibility, such as product and tort liability[25]. Considering the above-mentioned complications and intricacies this study on self-driving AVs and various case laws on self-driving car crashes and liability is important and significant and this will help in the drafting of AV laws for India.

1.6 Mens Rea and Actus Reus

The act (actus reus) requirement is typically the general premise for criminal responsibility. Only human behaviour qualifies as grounds for punishment[26]. It follows therefore that the actus reus and mens rea requirements for criminal culpability must be met for an AI's crime to be assignable to a human. To analyse the actus reus component of AI, it is vital to identify the humans participating in its decision-making[27].

When an AI commits a crime, the user is the main actor. The user is the one who first starts the AI, assigns it tasks, and reaps the rewards of the AI's labour. In criminal investigations into the behaviour of AIs, it has been the common practice thus far to target both the user and the supervisor[28]. The supervisor is the next potential actor; this person watches over the AI and has the power to influence its decisions if necessary. The maker of the AI develops it and maybe in control of all its components, including its hardware, software, and other features[29]. The producer is also knowledgeable about the technology that drives the AI's judgments while exposing it to the markets.

---

[23] The Basic Models of Criminal Liability of AI Systems and Outer Circles [online] available at https://papers.ssrn.com/sol3/papers.cfm?abstract_id=3402527 (last visited February 19, 2022)

[24] Artificial Intelligence and the External element of the Crime [online] available at http://www.diva-portal.org/smash/get/diva2:1115160/fulltext01.pdf (last visited February 19, 2022)

[25] Id.

[26] Id.

[27] Id.

[28] Id.

[29] Id.

The only actor who has the potential to change how the other actors perceive what the AI is capable of is the producer[30].

Even if the software developer may be a third-party functioning on the producer's behalf, the developer is considered to be a component of the producer for this study. In almost every situation, the owner will also be the user or the supervisor, and before a sale, the owner will also be the producer[31]. Despite this, the owner must be mentioned because they play a crucial part in the de lege ferenda study[32]. De Lege Ferenda means proposing an idea that could seek alteration of existing law through an amendment or repeal of existing law and execution of new law.

Finally, the outsider is a third party who does not directly communicate with the AI but who yet can influence its behaviour. That might be a person, such as a hacker or other similar individual, or in extreme circumstances, malware or a virus of some sort[33]. Impact and aptitude will primarily establish responsibility as factors for determining what caused the crime's effects, but they may also be crucial for determining criminal liability[34]. All the above scenarios apply to self-driving AVs too.

As of date in most jurisdictions, when we look at liability for crimes resulting from AI, an AI cannot be held legally responsible for its actions[35]. The legal requirement is to identify the human being who was using the AI in question. That person needs to be in a position where he or she might potentially have an impact on the AI and how it behaves. This appears to be something that will be decided after taking into account the particulars of each alleged act[36]. These actors could influence the AI by giving it precise instructions, remotely controlling it, or by failing to step in and override the AI's decisions. One can tell a drone to fly precisely while it is in the air or remotely control a drone to purposefully fly it into an aeroplane[37]. If one views the drone as a straightforward tool employed to harm the aeroplane from a responsibility perspective, the first case is not that challenging to resolve[38]. But

---

30. Id.

31. Id.

32. Feenan, D., 2013. Exploring the 'socio' of socio-legal studies.

33. Artificial Intelligence and the External element of the Crime, Supra Note 24, at 7.

34. Id.

35. Id.

36. Id.

37. Id.

38. Id.

when the drone takes control of itself, disregards the user's commands, and causes significant harm, the issue of culpability becomes more challenging. The question that would then come up would be, is it the user's or the manager's responsibility to stop the drone from acting in a certain way[39]? All actions attributed to a drone in the preceding few lines can also be attributed to a self-driving AV.

The primary areas of responsibility for the producer are the AI's hardware and software, which cover everything from mechanical components to internal code and algorithms, as well as the AI's education and training. Since the AI's code functions as its brain, centre, and the source of all its abilities, the producer can affect the AI in any way[40]. A malfunction that results from an at-fault producer will likely be linked to that manufacturer and its agents. However, what if AI, unintentionally commits crimes that the programmers were unaware of when they developed the code? Do these issues need to be resolved by the user or the supervisor[41]? Although the owner generally corresponds with the user or supervisor, it is advisable to exclude the owner as a potential defendant at this point in the analysis even though they are related. In actuality, the owner will be one of the other crucial players in the future of liability[42].

Criminal responsibility is obvious for anyone outside the AI who, in some way, affects its decision-making[43]. The outsider might be a hacker with the ability to modify the AI's programming or remotely manipulate it to ensure specific behaviour from the AI. The hacker may also be a third person who approaches the AI with a suggestion. Another possibility is for an outsider to deceive the AI and cause it to misinterpret its surroundings, leading to misbehaviour[44].

Liability for any kind of accident or incident which concerns self-driving AVs is part of the legal regime which is still under development. When you look at self-driving AVs it is very clear that the responsibility of driving would be with the machine than with a human as in a conventional automobile and

---

[39.] Id.
[40.] Id.
[41.] Id.
[42.] Id.
[43.] Id.
[44.] Id.

that there is a need for the current liability laws to be modified to come up with the right kind of remedy for any self-driving AV caused accident/damage.

The concept and usage of AVs have generated a lot of interest recently and basically, we can break down the issues into four problems.

The **first problem** is determining the necessary legal framework for permitting the testing and usage of Autonomous Vehicles. This could be in terms of regulations and rules for closed/open area testing, eligibility criteria to obtain Autonomous Vehicle Driving License, kinds of safety, product liability, servicing by manufacturer/manufacturer approved third party dealers and recording systems to be in place to facilitate investigation in case of accidents.

The **second problem** is to delve into required legal provisions for recording all the data concerning an Autonomous Vehicle through a data recorder (collection of evidence) starting from routine servicing, software updates, encryption of data in the recorder, listing authorised users of the recorded data, applicable data protection rights and principles for the usage of such recorded data and Information Technology laws with special reference to cyber security and hacking

The **third problem** is to identify the requisite "Legal Rights" and "Liability Regime" before permitting Autonomous Vehicles to be tested and used on public roads.

The **fourth problem** is to identify the required changes in the laws of India considering the above three problems to permit testing and usage of Autonomous Vehicles on public roads in India.

1.7　Existing Worldwide Legal Situation

The first death by an artificial person was caused by a Robot of Kenji Uradha, a 37-year-old Japanese engineer at the Akashai plant[45]. As per reports, the robot's work arm caught Kenji Uradha, trapping him against a device that breaks gears. He had repaired the robot in an off-limits area[46]. According to factory officials, the robot was surrounded by a wire mesh fence, which when unhooked would have cut off the electricity to the device. But Uradha

---

[45] Robot kills factory worker [online] available at https://www.theguardian.com/theguardian/2014/dec/09/robot-kills-factory-worker (last visited February 21, 2022).

[46] Id

appears to have leapt over the fence rather than opening it[47]. When the worker mistakenly touched the machine's on-switch while setting it to manual control, the robot's claw slammed him up against the machine's tooling. The robot's actions could not be stopped by other employees[48].

Then and even now various laws around the world are not properly equipped with a proper legal framework to deal with such situations. This problem has compounded with the development of AVs powered by AI and various countries are now working on laws to deal with AV regimes.

India as of date has no special law or provisions in the existing Motor Vehicle Act or allied Acts for testing and plying of AVs. Also "Product Liability" was loosely inferred from the old Monopolies and Restrictive Trade Practices (MRTP) Act 1969 (Now Competition Act 2002), the Sale of Goods Act 1930, and the old Consumer Protection Act 1986. However, under the new Consumer Protection Act 2019[49] or CPA 2019, Product Liability has been clearly defined under Section 2(34)[50] as the obligation of the producer or provider of any good or service to make up for any flaws and/or provide adequate remedies to the consumer[51].

If we permit AVs to ply in India without adding or altering the legal provisions in the existing laws and an accident occurs, then even though the courts would attempt to arrive at professional liability/culpability using the same traditional legal method, the litigation and sentencing would be a complicated one. The current procedure would be to rely on existing laws for civil and criminal liability, demonstrate the guilt or blameworthiness of each party involved in an AV accident, and then determine the level of responsibility and penalty. In short, in the absence of a sufficient, independent legal framework for offences/crimes committed by different participants in an AV accident, Courts would use the convoluted process of legacy rules to apportion blame in each instance.

[47.] Id.

[48.] Id

[49.] Product Liability Law in India – An Evolution [online] available at https://www.mondaq.com/india/dodd-frank-consumer-protection-act/974270/product-liability-law-in-india-an-evolution (last visited February 19, 2022).

[50.] Consumer Protection Act 2019 [online] available at https://egazette.nic.in/WriteReadData/2019/210422.pdf (last visited February 19, 2022).

[51.] Id

Some of the actors whose names would crop up in an Autonomous Vehicle accident as mentioned earlier would be the manufacturer of the AV, software coder, quality control team, driver/owner/user of the AV, the person responsible to provide updates for software, patches for malware and third-party hackers. This question of shared control and shared responsibility between humans and computers will come in all such AV-related cases and the prosecution will have to establish who is at fault, the human or the software.

Self-driving AVs will soon be employed for both personal and public transportation. Hence the government will have to come out with laws that can take care of potential issues in areas of cybersecurity, public safety, and privacy. Standards being introduced in these three areas will have to be aligned with the latest technologies. Also, new offences must be created to handle hackers who target such self-driving AVs. Laws must also be crafted to take care of pedestrians and cyclists who are not using AV technology. One important thing that must be kept in mind while making/amending existing laws to facilitate AVs is that the new laws must not impede innovation and development[52].

Presently many states in the USA including the District of Columbia have laws in place to regulate AVs. These laws started flowing in after the passing of the Federal Automated Vehicle Policy[53] in September 2016. Nevada state by passing Assembly bill 511 (36-6)[54] became the premier state in the USA to legalize driving AVs on the state highways of Nevada. Michigan Senate bill 995 on AVs was passed in 2018[55]. On 16 December 2019 California approved its revised regulations to allow the testing and plying of Autonomous delivery trucks with weights less than ten thousand and one pounds on San Jose

---

[52] Medium. 2021. *How Will State Regulations Fill the Gap for Self-Driving Cars.* [online] Available at: <https://mapanauta.medium.com/how-will-state-regulations-fill-the-gap-for-self-driving-cars-9030729e2c7f> [last visited November 8, 2021].

[53] Federal Automated Vehicle Policy [online] available at https://www.transportation.gov/AV/federal-automated-vehicles-policy-september-2016 (last visited March 23, 2021)

[54] Nevada Assembly Bill No 511 [online] available at https://www.leg.state.nv.us/Session/76th2011/Bills/AB/AB511_EN.pdf (last visited February 9, 2022). The bill defines "Artificial Intelligence", "Autonomous Vehicles" and "Sensors". The bill also lists the requisite standards for an Autonomous Vehicle before it can be operated on a highway in the State of Nevada.

[55] Michigan Assembly Bill No 995 [online] available at http://legislature.mi.gov/doc.aspx?2020-SB-0995 (last visited February 9, 2022)

public roads[56]. However, a permit from the Department of Motor Vehicles in California was mandatory before plying such AVs.

On 26 November 2018, Russia passed a regulation for the testing of AVs from December 2018 to March 2022[57]. According to Russian law, each driverless automobile must display a unique sign designating it as an autonomous vehicle, which includes a large black letter A. For the particular category of automobiles that are being tested, a pilot must have at least three years of professional driving experience, a spotless driving record, and medical certification (Para. 11)[58]. If no one else is found guilty, the owner of an AV is held accountable for all collisions with other vehicles and traffic incidents that happen on Russian roads. It is necessary to have liability insurance with a minimum of 10 million rubles (about $150,000) in coverage[59]. A legal entity may be the owner of a vehicle with an autonomous driving system; during the testing phase, the owner cannot be altered[60]. One of the salient features in the Russian regulation is the requirement of a mandatory Data Recording System[61], wherein the video (data) recording is required to be kept for at least 10 years and can be provided to government entities at their request[62].

Germany has a strategy for AVs at the National level. Their draft regulation was passed by the Cabinet on 10 February 2021 and sent to Bundestag (upper chamber) and Bundesrat (lower chamber). Bundestag passed the AV regulation on 20 May 2021 and Bundesrat on 28 May 2021[63]. Some of the salient features in German regulations are test procedures, cyber security, and approval of autonomous driving functions[64]. Canada has an existing framework for AVs,

---

[56] California Assembly Bill [online] available at https://www.dmv.ca.gov/portal/vehicle-industry-services/autonomous-vehicles/california-autonomous-vehicle-regulations/#:~:text=On%20December%2016%2C%202019%2C%20the,an%20approved%20permit%20from%20DMV) (last visited, February 19, 2022)

[57] Russia: Government Begins Testing Driverless Cars [online] available at https://www.loc.gov/item/global-legal-monitor/2019-01-18/russia-government-begins-testing-driverless-cars/ (last visited, February 19, 2022).

[58] Id.

[59] Id.

[60] Id.

[61] Id.

[62] Id.

[63] Germany first country to pass AV regulations [online] available at https://www.eetasia.com/germany-first-to-pass-av-regulations/ (last visited February 19, 2022)

[64] Strategy for Automated and Connected Driving – BMVI [online] available at https://www.bmvi.de/SharedDocs/EN/publications/strategy-for-automated-and-connected-driving.pdf?__blob=publicationFile (last visited February 19, 2022)

but it does not permit an AV without steering and pedals to be driven on the roads as per their existing Motor Vehicle Safety Act[65].

China first permitted road tests for AVs in Beijing in December 2017. This was followed by Shanghai in March 2018 and the country-wide national-level regulations for a road test of AVs in April 2019. China is slowly and steadily working on AVs and tweaking the existing laws to support its deadline for mass production of AVs by 2025[66]. As per Chinese law, a system that is capable of performing autonomous driving tasks for a certain amount of time is referred to as an autonomous driving system. The automated driving system does not include driver assistance features like electronic lane-keeping assistance, lane departure warning, parking assistance, emergency braking, adaptive cruise control, electronic blind spot monitoring, and traffic congestion queuing support[67]. Australia, the United Kingdom (UK) and New Zealand are working on appropriate frameworks/regulations for AVs.

Worldwide, "Product Liability" points to the legal liability of a manufacturer or seller who sells a defective product to a consumer. In the USA Product Liability varies between the states. However, one thing is common i.e., there is no cap on the liability in any state. The EU Product Liability Directive was signed in 1985 and implemented in 1988[68]. This Directive contains many exculpation clauses for the producer of which one is proof of not putting the defective product into circulation. This directive is prospective, and the manufacturer compensation (per defect) is capped at Euro (EUR) 70 million. Australia, Taiwan, China, and Japan also have Product Liability laws. Japan has a time limit of 3 years from the time and date of injury to make a compensation claim. Also, under the Japan law, manufacturers cannot be held responsible if product delivery was before ten years[69]. Like Japanese law, Chinese laws also have time limitations.

---

[65] Canada's Safety Framework for automated and connected vehicles [online] available at https://tc.canada.ca/sites/default/files/migrated/tc_safety_framework_for_acv_s.pdf (last visited February 19, 2022).

[66] Beijing issued first rules on road tests for Autonomous Vehicles [online] available at https://www.globalcompliancenews.com/2018/01/22/china-autonomous-vehicles-20180122/ (last visited February 19, 2022).

[67] Id

[68] European Agency for Safety and Health at Work [online] available at https://osha.europa.eu/en/legislation/directives/council-directive-85-374-eec (last visited February 19, 2022).

[69] Japan Product Liability Laws and Regulations [online] available at https://iclg.com/practice-areas/product-liability-laws-and-regulations/japan (last visited February 19,2022).

CHAPTER 2

LEGAL REGIME IN USA/ MAJOR EUROPE NATIONS FOR AUTONOMOUS VEHICLES

2.1 Background of Autonomous Vehicles in the United States of America (USA) and Major Europe Nations

The history of autonomous cars in the United States of America ("USA") finds its roots in a 1939 exhibition at New York World's Fair. General Motors ("GM") for that exhibition had created an exhibit named Futurama to display their vision of the world twenty years later. This exhibit included smart highway systems and AVs[70]. The first autonomous electric vehicle was developed by Norman Bel Geddes. Using magnetised metal spikes buried in the road, radio-controlled electromagnetic fields were used to direct this vehicle[71]. General Motors brought this idea to life around 1958. The car's front end was equipped with pick-up coils, and sensors to measure the current passing through a roadside wire[72]. By changing the current, the steering wheel of the car could be moved to the left or the right. The Japanese expanded on this concept in 1977 by utilising a camera system that transmitted data

---

[70] History of the Autonomous Car, TitleMax (2020), https://www.titlemax.com/resources/history-of-the-autonomous-car/#:~:text=History of Autonomous Cars, made this concept a reality (last visited January 10, 2021).

[71] Id.

[72] Id.

to a computer so it could process photographs of the route[73]. However, the maximum speed of this vehicle was less than 20 mph. The Germans improved a decade later with the VaMoRs, a camera-equipped vehicle that could safely travel 56 mph[74]. The capability of self-driving cars to notice and respond to their surroundings has evolved along with technology. Although autonomous cars are only still in the testing stages in many parts of the world, we know that the world for sure is on track to reach there[75].

The innovation process of autonomous cars has resulted in the advent of many vehicles which are presently running on the road and these vehicles can be called semi-AVs. These vehicles have many features including assisted parking systems. But the presence of a human driver is mandatory in these cars, thereby necessitating no changes in present laws to permit these vehicles to run on the road. Many advantages are anticipated from AVs, including a decrease in accidents caused by drunk driving and a ban on distracted driving while using a cell phone. In short, all accidents that happen now which are attributable to human error can be completely controlled by the use of AVs. The expectation in the USA is that this technology will progress and by 2025 at least half of the vehicles rolling out of assembly lines in the automobile industry would be AVs.

Up to November 2020, thirty-eight states in the USA along with DC have passed laws/executive orders for testing and plying of AVs[76], Sixteen states and the District of Columbia permitted full deployment, while five states authorised a study to clarify essential words, state contacts, or funding. Twelve states also authorised testing[77]. Among them, 18 states currently permit testing or deployment without a human operator in the car, while 4 states have laws that govern truck platooning[78]. Other states either have them in the pipeline or are yet to decide on them.

AVs are rated according to their automation levels. The basic standard is six levels of automation. All manufacturers are mandated to carry out safety assessments as per Department of Transport, USA norms. Irrespective of

---

[73.] Id.

[74.] Id.

[75.] Id.

[76.] Autonomous Vehicles, GHSA, https://www.ghsa.org/taxonomy/term/521 (last visited January 10, 2021).

[77.] Id.

[78.] Id.

the level of automation there is no excuse for obtaining vehicle registration, insurance, and driving license.

Different levels of automation are categorised using a zero–five rating scale the Society of Automotive Engineers developed, in the USA (SAE). Greater automation means a higher level and lower automation means a lower level[79]. The various levels are:

- **Level Zero: No Automation** – The driver is solely accountable.

- **Level One: Driver Assistance** – The vehicle is programmed to do partial work like say cruise control. However, a driver has to be present and must perform assigned tasks.

- **Level Two: Partial Automation** – Help is provided to the driver by the vehicle on one or more systems. Driver's presence is mandatory.

- **Level Three: Conditional Automation** – AV is automated in all aspects, but the presence and intervention of a driver are required when necessary

- **Level Four: High Automation** – The vehicle is fully automated and can perform all duties even if driver intervention is not there. However, the driver's presence is mandatory.

- **Level Five: Full Automation** – Fully automated driverless vehicle.

When one talks of Europe and automobiles, one country that rings a bell is Germany. There are so many world-famous automobile companies from Germany like Volkswagen, Mercedes, and BMW. As expected, Germany has also been the pioneer nation in Europe to introduce an Autonomous Vehicle or AV policy. Germany enacted a law in 2017, which permitted AVs to be tested and operated on public roads[80]. According to the German Autonomous Driving Act, a motor vehicle that meets certain technical requirements and is capable of performing the driving task autonomously within a specific operating region is considered to have autonomous driving capabilities[81]. However, this was subjected to certain conditions. One of them was the

---

[79] SAE Levels of Driving Automation [online] available at https://www.sae.org/blog/sae-j3016-update (last visited, February 22, 2022)

[80] Germany: Road Traffic Act Amendment allows driverless vehicles on public roads [online] available at https://www.loc.gov/item/global-legal-monitor/2021-08-09/germany-road-traffic-act-amendment-allows-driverless-vehicles-on-public-roads/ (last visited March 02, 2021.

[81] Id.

mandatory requirement to have a safety driver in the car[82] who could control the car and the steering wheel on a required basis. The present legislation though does not legally recognize level five or full AVs, it has a construct to provide adequate opportunities for testing of AVs within the country of Germany. According to the German definition of a defined operating area (Autonomous Driving Act, Art. 1, No. 1, 1d), the same applies to the locally and physically established public road area where an autonomous vehicle may be operated as permitted by the appropriate state body[83].

The German regulation of 2017 apart from creating a roadmap for AVs also worked to solve liability issues[84]. Due to the inherent conflict of interest, there are certain tensions between the severe liability imposed on a car manufacturer under Section 1 of the Product Liability Act and the vehicle owner under Section 7 of the Road Traffic Act. The automobile keeper notion no longer corresponds to the shift in society away from an ownership mentality and in the direction of a sharing community with a platform, and the maker can easily avoid the product obligation[85]. Some of the major issues and questions which were dissected in this regulation were, Can the driver be earmarked as a responsible person in an AV?; If the AV is jointly being controlled by humans and machines, what should be the process to decide on civil and criminal liability/responsibility[86]?; How effectively can AVs be used in public transport areas?; What is the forecasted demand or likely demand for driverless passenger transport?; The trend, as seen in many other industries, is away from selling products and toward providing services. Mobility as a Service, or MaaS, refers to the practice of leasing cars[87]. Can AVs be car shared through new business models like MaaS (Mobility as a Service)[88]?; Is it necessary to create new criminal offences to address the various forms of behaviour that could involve the usage of AVs? Impact of using AVs on road users and ways and means to protect road users from associated risk.

---

[82] Id.

[83] Id.

[84] Tilburg University AVs Regulation in Germany and the US and its impact on the German car industry [online] available at http://arno.uvt.nl/show.cgi?fid=149595 (last visited March 02, 2021).

[85] Id.

[86] Id.

[87] Id.

[88] Id.

Around May 2018, the European Commission published a report[89] which was titled "On the road to automated mobility: An EU strategy for mobility of the future"[90]. This document listed out a roadmap to outline the ambition of the EU towards becoming a world leader in deploying AVs. The communication released by the EU Commission talked about the allocation of budget and investments towards the development of technology and requisite infrastructure for the operation of AVs, the Development of demand in local markets to accept the usage of AVs, Issuance of detailed guidelines for plying of AVs including various road related safety aspects and features for AVs. Liability aspects connected with accidents and also covering cybersecurity, data protection, and access and Study on impact and implications of AVs on society and economy as a whole and to analyse the negative impacts if any needing changes to current legislation or introduction of new laws. The Commission was of the firm view that the existing EU legislation was broadly conducive to permitting AVs to be put on the road and that only some changes were required in the law[91] to synchronize the plying of AVs and non-AVs as a process towards making the law future-proof. The communication released by the EU Commission also talked about liability aspects connected with accidents and also covering cybersecurity, data protection and access, a study on the impact and implications of AVs on the society and economy as a whole and to analyse negative impacts if any needing amendments to existing laws or introduction of new laws.

The Commission allocated an amount of Euro 450 Mn under its flagship "Connecting Europe Facility" program to help support the introduction of AVs through digitization[92]. As per the article, the Connecting Europe Facility (CEF) is an important EU financial tool for promoting economic development, job creation, and competitiveness at the European level. It promotes the development of highly efficient, ecologically friendly, and seamlessly linked trans-European networks for energy, digital services, and transportation. Europe's energy, transportation, and digital infrastructure are

---

[89] EU COM (2018) 283 final. On the road to automated mobility: An EU strategy for mobility of the future [online], https://eur-lex.europa.eu/legal-content/EN/TXT/?uri=CELEX%3A52018DC0283 (last visited March 02, 2021).

[90] Id.

[91] Id.

[92] Connecting Europe Facility [online] available at https://ec.europa.eu/inea/en/connecting-europe-facility (last visited March 02, 2021).

completed by CEF investments[93]. The Commission has encouraged all EU member countries and regions to take advantage of this allocated funding and appropriately use them for research and development of AVs and associated technologies. The Commission has also requested member states to finalise and prioritize their list of use cases for large-scale testing of AVs. These countries have also been advised to be mindful of likely synergies between connectivity and automation use cases. The Commission has also promised that it will ensure the availability of a common platform in the EU where all relevant public and private stakeholders will be mapped to facilitate each one of them to coordinate road testing of AVs and also discuss and work towards solving pre-deployment issues.

The Commission has also committed to supporting the existing work in this field at the EU and national levels. The Commission has also agreed to offer free worldwide high-accuracy services through the satellite Galileo and has committed to preparing appropriate guidelines for use of satnav systems like EGNOS and Galileo[94]. Post this agreement by the commission, satnav systems can be included in vehicle navigation systems, and this will help address safety and liability issues.

During the year 2017, an Ethics Committee was formulated in Germany to study and make recommendations on ethical programming in AVs[95]. This committee came up with a report containing 20 different principles. Some of these principles included aspects concerning the usage of networked AVs.

The phrase "connected driving" was chosen to draw attention to the ethical issues that the committee was also debating about the networking and informational linkage of automobiles. As per these principles, if there is an imminent danger, then the protection of human life[96] must take priority, qualification of people as per their characteristics like sex and age should not be done in cases of unavoidable accidents, and when it comes to AVs, basis the driving conditions and situations the law must clearly define the

---

[93] Id.

[94] Green light for ESA's advanced satnav technology and innovation programme https://www.esa.int/Applications/Navigation/Green_light_for_ESA_s_advanced_satnav_technology_and_innovation_programme (last visited March 02, 2021).

[95] The German Ethics Code for Automated and Connected Driving [online] available at https://www.researchgate.net/publication/320011270_The_German_Ethics_Code_for_Automated_and_Connected_Driving (last visited March 02, 2021).

[96] Id.

parameters to identify the person/machine responsible for driving task as this will kill any kind of ambiguity. Lastly, the principles also stated that any personal information collected from the AVs must be done with the consent of the safety driver and they must have a say in what data can be collected. Levels 4 and 5 of the German system are combined into one by the National Highway Traffic Safety Administration (NHTSA). The German system led the committee to believe that levels 4 and 5 should be their primary focus, even though they haven't been realised yet. Thus, full driving automation and driverless cars were at the centre of deliberation.

2.2 Necessity of AVs

Statistics from the National Safety Council (NSC) reveal that around 40,000 plus people are killed every year due to road accidents in the USA. The deaths due to such motor vehicle accidents have been increasing year after year[97]. Roadways in the USA continue to present some of the greatest risks every day, whether one is travelling by car, motorcycle, or even just crossing the street[98]. In addition to these deaths, millions of injuries have occurred due to road accidents and the medical attention costs in the USA have also been close to millions of dollars per year[99]. The NSC predicts that around 21,340 people may have died on USA roads in the first half of 2022[100]. To ensure travelers get to their destination as securely as possible, NSC urges all road users to adhere to safe driving guidelines[101]. The National Highway Traffic Safety Administration (NHTSA) carried out a study and it proved that human error has been the reason for more than 90% of the deaths caused due to road accidents.[102] More than a quarter (28%) of all fatalities nationwide occurred in collisions involving drunk drivers. The number of fatalities caused by drunk

---

[97] Protect Yourself and Loved Ones by Addressing Roadway Risks [online] available at https://www.nsc.org/road-safety/safety-topics/fatality-estimates, (last visited February 23, 2022). The National Safety Council estimates over 42,000 people were killed in motor vehicle crashes in 2020, an 8% increase over 2019, and the first half of 2021 is shaping up similarly.

[98] Id.

[99] Id. NSC has released preliminary estimates that show a 16% increase in motor vehicle deaths from the first six months of 2021.

[100] Id.

[101] Id.

[102] 2016 Fatal Motor Vehicle Crashes: Overview [online] available at https://crashstats.nhtsa.dot.gov/Api/Public/ViewPublication/812456 (last visited February 23, 2022).

driving decreased in 17 States and Puerto Rico. Mississippi saw the biggest drop, with 49 fewer people dying in accidents involving drunk drivers in 2016[103]. The number of fatalities involving alcohol-impaired driving increased in 29 States plus the District of Columbia, with California experiencing the highest increase of 148 fatalities followed by Florida with 53 more[104]. The NHTSA then concluded that an AV that can be programmed to drive on its own without any kind of error was probably the best solution to put an end to the huge number of humans being killed by humans each year in the USA[105]. It would also greatly minimise the massive emotional toll on families, in addition to minimising the enormous social costs involved with crashes in terms of lives lost, hospital stays, missed days at work, and property damage, which amount more than hundreds of billions of dollars annually[106]. Additionally, these profound shifts will present a wealth of new prospects for investments in the underlying technology as well as employment across a range of industries that create, produce, and maintain them. Insurance companies are also supporting this, and they have started offering discounts to drivers of vehicles with safety devices.

2.3 Ethical Issues to be considered while programming AVs and drafting changes to existing laws

In current road accidents, human drivers make choices on what to avoid or rather what to save and in most of these cases, these decisions are founded on the culture of that country/state and may not be uniform across the world. When we start having full AVs these vehicles will have to make choices on what to be saved or avoided and this will eventually depend on how these vehicles have been programmed. Human bias or human ethics based on the region where these vehicles are manufactured can be one strong factor in deciding how these vehicles are to be programmed. But this may not work, or the automobiles might not always be popular across the world unless they are programmed separately basis the culture and laws/judicial precedents of the

---

[103]. Id.

[104]. Id.

[105]. National Highway Traffic Safety Administration: Preliminary [online] available at https://www.nhtsa.gov/staticfiles/rulemaking/pdf/Automated_Vehicles_Policy.pdf (last visited February 23, 2022).

[106]. Id.

places where these vehicles are to be sold or operated. Massachusetts Institute of Technology (MIT) launched a survey[107] called Moral Machine and collected lakhs of responses across the world on choices humans would make in terms of whom to save, when it comes to accidents. Some of the findings were that humans would save "Young over old", "Humans over animals", and "Few over many"[108]. Most of them were not sure when it came to choosing between senior citizens and pregnant ladies.

As of date, it is a very difficult proposition to even think of programming a vehicle to take such ethical decisions. However, the day is not far when AVs will be the in-thing on the road and at that stage, AVs will have to take these ethical decisions. Hence programming AVs to make ethical decisions is a subject that must be dealt with keeping in mind the culture and judicial precedents of the places where these vehicles operate.

2.4 Existing laws, regulations, and policies on AVs in the USA

The USA Department of Transportation (DOT) has come up with specific guidelines[109] for manufacturers involved in the creation and evaluation of vehicles. The federal responsibilities listed in the guidelines are Setting Safety Standards and enforcing compliance, investigating non-compliance, managing recall of faulty automobiles, promoting automobile safety among the general public, and issuing regulations to meet national objectives in all areas concerning automated vehicles including safety and compliance. The fundamental premise of the policy is the DOT's view that AVs can help reduce accidents through better safety, sustainability, and mobility[110]. The National Highway Traffic Safety Administration (NHTSA) published its first version of the Federal Automated Vehicles Policy (FAVP) in September

---

[107] Peter Dizikes | MIT News Office, how should AVs be programmed? MIT News | Massachusetts Institute of Technology, https://news.mit.edu/2018/how-autonomous-vehicles-programmed-1024 (last visited January 10, 2021).

[108] Id.

[109] Fact Sheet: Federal Automated Vehicles Policy Overview, https://www.transportation.gov/sites/dot.gov/files/docs/DOT_AV_Policy.pdf (last visited January 10, 2021).

[110] Id. The state responsibilities listed in the guidelines are Human Driver's License, Registration of Automated Vehicles in their jurisdiction, Enactment of Traffic Laws and their enforcement, carrying out safety inspections on as required basis and Regulating insurance and liability of such automated vehicles.

2016[111]. The federal government is still in charge of establishing the minimum safety requirements for motor vehicles, even though states continue to be the principal regulators when it comes to licensing, registration, traffic law enforcement, safety inspections, infrastructure, insurance, and liability. For states that desire to proceed with testing and ultimately implement AVs, the MSP provides a roadmap[112].

As per Section 2, the Model State Policy (MSP) describes federal versus state authority[113]. Instead of providing a comprehensive set of legislative texts, the FAVP proposes actions a state could take. Even while the organisation might use future regulations to make some of the advice necessary and binding, it specifically states that this guideline is not mandatory. Additionally, it points out several state laws that could need to be updated to account for a world full of automated vehicles[114]. Some of these include emergency response and law enforcement, vehicle registration, liability and insurance, education and training, vehicle inspections and maintenance, and environmental effects[115].

The Consolidated Appropriations Act, 2018 was adopted as legislation on March 23, 2018. As per this Act, U.S. DOT is to carry out research into the creation of automated vehicles, especially in the domain of Highly Automated Vehicles ("HAV") technologies and Advanced Driver-Assistance Systems (ADAS)[116]. A nonpartisan, bicameral measure about ADAS technology and AVs has been created by the House Committee on Energy and Commerce and the Senate Committee on Commerce, Science, and Transportation[117].

2.5 Self-Drive Act and AV Start Bill

Numerous states started to draft their state policies on AVs resulting in differential policies across the states, complicating interstate travel. The

---

[111.] Anne Teigen Ben Husch, REGULATING AVs, https://www.ncsl.org/research/transportation/regulating-autonomous-vehicles.aspx (last visited January 10, 2021).

[112.] Id.

[113.] Id.

[114.] Id.

[115.] Id.

[116.] USDOT Comprehensive Management Plan for Automated Vehicle Initiatives, U.S. Department of Transportation, https://www.transportation.gov/policy-initiatives/automated-vehicles/usdot-comprehensive-management-plan-automated-vehicle (last visited December 17, 2020).

[117.] Id.

varying policies also deter innovation and development of AVs[118]. Since then, the National Highway Traffic Safety Administration (NHTSA) has continued to look for methods to lower regulatory obstacles to these technologies as the automobile industry has continued to make advancements in the development of AVs and automated driving systems (ADS)[119]. To move ahead in the development of AVs in 2017 the US House of Representatives passed the Safely Ensuring Lives Future Deployment and Research in Vehicle Evolution Act (SELF DRIVE Act) vide H.R. 3388[120]. A nonpartisan, bicameral law about ADS technology and AVs has been developed by the House Energy and Commerce Committee and the Senate Committee on Commerce, Science, and Transportation. The bill attempted to clarify federal and state governments' functions regarding this emerging technology and was primarily aimed at adopting federal safety standards in the arena of AVs. Post passing of this Act, a bill like that called American Vision for Safer Transportation through Advancement of Revolutionary Technologies Act (AV START Act) under reference S. 1885 was also presented but never passed[121]. This bill establishes a framework for the federal government's responsibility for the safety of highly automated vehicles (HAVs), forbids states from adopting, upholding, or enforcing laws, rules, or standards that govern HAVs or automated driving systems (ADSs), outlines the requirements for the testing, assessment, or demonstration of HAVs in interstate commerce and requests several safety exemptions[122].

The draft language of the AV Start Act is quite similar to the SELF-DRIVE Act and calls for the creation of a Highly Automated Vehicle Advisory Council within the Department of Transportation to assess issues relating to AVs[123]. It also addresses federal laws surrounding the testing of AVs[124]. The goals of this bill were to: (1) establish a framework for the federal government's

---

[118] Congress Taking Another Look at Regulating Automated Driving Systems, JD Supra, https://www.jdsupra.com/legalnews/congress-taking-another-look-at-51054/ (last visited December 17, 2020).

[119] Id.

[120] Id.

[121] S 1885 – AV Start Act [online available at https://www.congress.gov/bill/115th-congress/senate-bill/1885#:~.text=This%20bill%3A%20(1)%20establishes,report%20subject%20areas%3B%20(3) (last visited February 23, 2022).

[122] Internet of Things, Artificial Intelligence and Smart Dust Technologies, https://www.internetlawyer-blog.com/internet-of-things-artificial-intelligence-and-smart-dust-technologies/ (last visited September 19, 2022).

[123] S 1885 – AV Start Act, Supra Note 141, at 44.

[124] Id.

responsibility for ensuring the safety of highly automated vehicles (HAVs); (2) stop states from adopting, maintaining, or enforcing any laws, rules, or standards regulating an HAV or automated driving system (ADS) regarding certain safety evaluation report subject areas; and (3) specify the circumstances under which HAVs may be introduced into interstate commerce for testing, evaluation, or demonstration[125].

Under the AV Start Act, the onus was vested in each AV manufacturer to provide a defined strategy for assessing and minimising cybersecurity risks. The Department of Commerce was then responsible for setting up a committee for data access that would make appropriate policy recommendations concerning data production and sharing. The Self-Drive Act also had certain great features. It clarified the role of NHTSA in regulating the safety, design, and construction of AVs. It improved the access of NHTSA to safety data. This was done with the purpose that NHTSA could plan any required upgrades and research and development (R&D)[126]. The Self-Drive Act created NHTSA safety evaluation certifications for manufacturers and improved protection for consumer education, cybersecurity, and privacy.

The main disadvantage the Self-Drive Act had was the absence of regulatory or testing requirements[127]. To speed up the release of these automobiles onto the market, the Self-drive legislation also proposed to exclude the first 100,000 vehicles produced by a manufacturer from certain existing safety requirements. According to experts, the Self-Drive Act shields technology and automakers from liability when their unregulated technologies endanger users. Finally, both bills came to an end when the new Congress took office in 2018. The Self-Drive Act, also known as the Safely Ensuring Lives Future Deployment and Research in Vehicle Evolution Act was reintroduced on 23 September 2020, by Reps. Greg Walden (R-Ore.) and Bob Latta (R-Ohio) to establish what they refer to as a national framework for the development and testing of AVs[128].

---

125. Id.

126. H.R. 3388 (115th): Self Drive Act, Govtrack.us, https://www.govtrack.us/congress/bills/115/hr3388/summary (last visited December 17, 2020).

127. Id. The Self-drive act also planned to grant exemptions to existing safety standards for a company's first 100,000 vehicles, to speed up their production into the market

128. Republican House Members Reintroduce Self Drive Act, AASHTO Journal (2020), https://aashtojournal.org/2020/09/25/republican-house-members-reintroduce-self-drive-act/ (last visited December 17, 2020). Rep. Greg Walden, R-Ore., and Rep. Bob Latta, R-Ohio, re-introduced the Safely

2.6   History of legislation on AVs in various States in the USA[129]

As seen earlier USA has many states which have enacted laws in detail to permit AVs to test and ply in their states. We shall study the law in various states in the USA under three buckets viz individual state laws where more detailed laws exist, collective study of laws in states where AV laws are in developing stages, and finally laws in states which have executive orders.

2.6.1   *California*

Vide Senate Bill No. 1298 Chapter 570[130], approved by the Governor on September 25, 2012, the Department of Highway Patrol in California was mandated to set up safety standards and other requirements to facilitate the testing of AVs on its public roads. To undertake such tests the AVs were also required to meet the outlined standards and requirements[131]. The following definitions have been enumerated under Division 16.6 of said Bill:

"Autonomous technology means technology that has the capability to drive a vehicle without the active physical control or monitoring by a human operator"[132].

"Autonomous vehicle means any vehicle equipped with autonomous technology that has been integrated into that vehicle. An autonomous vehicle does not include a vehicle that is equipped with one or more collision avoidance systems, including, but not limited to, electronic blind spot assistance, automated emergency braking systems, park assist, adaptive cruise control, lane keep assist, lane departure warning, traffic jam and queuing assist, or other similar systems that enhance safety or provide driver assistance, but are not capable, collectively or singularly, of driving the vehicle without the active control or monitoring of a human operator"[133].

Ensuring Lives Future Deployment and Research in Vehicle Evolution or Self Drive Act on September 23 to create what they call a "national framework" for the development and testing of AVs or AVs.

[129] Melanie Musson Published Insurance Expert, which states allow self-driving cars? (2020 update) autoinsurance.org | compare auto insurance companies, search 100's of auto insurance reviews, and get cheap auto insurance quotes online. save $ $ $ on your auto insurance rates! (2021), [online] available at https://www.autoinsurance.org/which-states-allow-automated-vehicles to drive on the road/#5 (last visited December 17, 2020).

[130] State of California Senate Bill No 1298, http://www.leginfo.ca.gov/pub/11-12/bill/sen/sb_1251-1300/sb_1298_bill_20120925_chaptered.pdf (last visited December 22, 2020).

[131] Id.

[132] Id.

[133] Id.

"An operator of an autonomous vehicle is the person who is seated in the driver's seat, or if there is no person in the driver's seat, causes the autonomous technology to engage"[134].

The bill also states that a driver with the appropriate class of license can test an AV on public roads.

Vide Assembly Bill No. 1592, passed on August 29, 2016, testing was allowed by Contra Costa Transportation Authority on AVs[135] that had no steering, driver, and any kind of pedals. However, the bill permitted testing only at specified places with a vehicle speed of 35 miles or below. As per the bill:

"A manufacturer of autonomous technology is the person as defined in Section 470 that originally manufactures a vehicle and equips autonomous technology on the originally completed vehicle or, in the case of a vehicle not originally equipped with autonomous technology by the vehicle manufacturer, the person that modifies the vehicle by installing autonomous technology to convert it to an autonomous vehicle after the vehicle was originally manufactured"[136] (Assembly Bill 1592).

Further, the bill also states that testing may take place at GoMentum Station, which is located inside the walls of the former Concord Naval Weapons Station, and at privately held industrial parks recognised by the authority, including any public roadways. The right to perform testing under this section shall not be affected by a change in ownership of the real estate that makes up the GoMentum Station. A surety bond of $5 million was to be submitted before commencing testing[137].

Vide Assembly Bill No. 669, Chapter 472, approved on 04 October 2017, California changed Section 14107 of the Government Code to enable the DOT to test technology with the Department of Highway Patrol so that drivers could safely operate motor vehicles with fewer than 100 feet between each vehicle or group of vehicles[138]. As per the bill the Department of Transportation may only conduct testing on roads, streets, and highways

---

[134.] Id.

[135.] Id.

[136.] Id.

[137.] Assembly Bill No 1592 [online] available at http://www.leginfo.ca.gov/pub/15-16/bill/asm/ab_1551-1600/ab_1592_bill_20160830_enrolled.pdf (last visited February 23, 2022).

[138.] Bill text - AB-669 Department of Transportation: Motor Vehicle Technology Testing., https://leginfo.legislature.ca.gov/faces/billNavClient.xhtml?bill_id=201720180AB669 (last visited December 22, 2020).

that the California Highway Patrol Department has approved. A person must own an active driver's license appropriate for the participating vehicle's class to operate an automobile that is being tested.

Vide Assembly Bill No. 1444, Chapter 719 approved on 12 October 2020, Section 38756 was added to the Vehicle Code by California, which permitted Livermore Amador Valley Transit Authority[139] to execute a shared autonomous vehicle (SAV) demonstration project for testing AVs without a driver in the driver's seat, a steering wheel, a brake pedal, or an accelerator under specified circumstances[140]. As per new Section 38756, only the Californian city of Dublin may be used for testing, and only within the demonstration project's boundaries may vehicles travel on public roadways. The driverless car must travel at speeds of no more than 35 mph. Before testing of an autonomous vehicle on or over a public road starts, the Livermore Amador Valley Transit Authority, a private organisation, or a combination of the two should complete both of the following: (1) Acquire a five-million-dollar ($5,000,000) instrument of insurance, surety bond, or proof of self-insurance, and provide the Agency of Motor Vehicles with documentation in the type and manner required by the department. (2) Send a full explanation of the testing programme to the department[141].

2.6.2 *Florida*

In July 2012 Florida declared that it would support the development, testing, and operation of AVs[142]. In 2012, Florida's first AV regulations included a testing phase and a report (the Report) from the Department of Highway Safety and Motor Vehicles (DHSMV). At that stage, the understanding was that a driver of an AV would require an appropriate driver's license and adequate insurance. DHSMV was given the assignment to write and submit a report on this subject. With effect from 01 July 2016, Florida approved

---

[139.] Bill Text - AB-1444 Livermore Amador Valley Transit Authority: demonstration project., https://leginfo.legislature.ca.gov/faces/billTextClient.xhtml?bill_id=201720180AB1444 (last visited December 22, 2020).

[140.] Id.

[141.] Id.

[142.] An Attempt to Control What Controls Itself: Unraveling Florida's Autonomous Vehicle Laws [online] available at https://www.phelps.com/a/web/6rLYXJMb3axzKRFn2GR53m/ta-vol-38-n3-2019-dahdah-av-laws.pdf (last visited February 23, 2022).

two laws that allowed drivers with valid licenses to operate AVs on public highways. Also, the AVs were required to maintain federal safety standards. A study on Autonomous truck platooning was also commenced.

On 01 July 2019, a revised bill on AVs was passed which did not mandate the presence of a driver in a fully autonomous vehicle.[143] As per 316.003 of the bill, the term any vehicle with an automated driving system is considered an autonomous vehicle. All real-time operational and tactical activities are referred to as dynamic driving tasks, if any, necessary to drive a vehicle in on-road traffic while operating within the parameters of its operational design. Strategic tasks like trip planning and destination and waypoint selection are not included. However, the bill also mandated the physical presence within the USA of a remote human operator with an appropriate valid USA driving license. The bill exempts AVs[144] from rendering aid. A completely autonomous vehicle, as defined by the bill, is a vehicle equipped with an automated driving system and designed to operate without a human driver[145].

The Florida Turnpike Enterprise, a division of the Florida Department of Transportation (FDOT), is permitted by the bill's section 338.2216 to enter into one or more agreements to develop cutting-edge, autonomous, networked transportation technologies; and to finance, build, and run facilities for their growth[146]. The bill mandates specified insurance coverage for full AVs with the ADS engaged. If the owner of a fully autonomous vehicle involved in an accident lacks the required insurance coverage, then as per the bill the owner must give documentation of their financial responsibility while responding to damages claim[147]. The legislation forbids local governments from levying any taxes, fees, and requirements on for-hire vehicles, or placing other requirements on AVs or on those who drive them[148]. These requirements will be valid till January 1, 2024. The bill expresses legislative intent to ensure that there is uniform law for AVs across the state[149]. The proposed legislation completely

---

[143] Florida House of Representative CS/HB 311, https://www.flsenate.gov/Session/Bill/2019/311/BillText/er/PDF (last visited December 24, 2020).

[144] Id.

[145] Id.

[146] Id.

[147] Florida Senate Committee of Infrastructure and Security cs/hb 311, https://www.flsenate.gov/PublishedContent/Session/2019/BillSummary/Infrastructure_IS0311is_0311.pdf (last visited December 24, 2020).

[148] Id.

[149] Id.

exempts AV operators from the requirements of Chapter 316 of the Florida Statutes, including the requirement to provide information and administer aid in the event of an accident. The same goes for any laws about unattended motor vehicles or other properties that are regarded to not apply to such fully AVs. The law adds exclusions for AVs to existing sections relating to wireless communications, the usage of visual displays, and other laws[150].

2.6.3 Georgia

On 01 July 2017 Georgia passed two bills. One bill exempted the prohibition of following too closely to a vehicle that isn't in the lead in a convoy travelling in the same lane and using vehicle-to-vehicle communication technology to coordinate their movements[151]. The Official Code of Georgia annotated in Chapter 40, which deals with traffic and motor vehicles, is changed in Code Section 40-6-49, relating to following too closely, by adding a new subsection to read as follows:

"(e) This Code section shall not apply to the operator of any non-leading vehicle travelling in a coordinated platoon. For purposes of this subsection, the term 'coordinated platoon' means a group of motor vehicles travelling in the same lane utilizing vehicle-to-vehicle communication technology to automatically coordinate the movement of such vehicles"[152] (Georgia HB No 472).

The second bill exempted operators of automated vehicles from having a driving license and permitted automated vehicles to be driven without a human in them[153]. This bill also in para 17.2 defines:

"Fully autonomous vehicle as a motor vehicle equipped with an ADS that has the capability to perform all aspects of the DYNAMIC driving task without a human driver within a limited or unlimited operational design domain and will not at any time request that a driver assume any portion of

---

150. Id.

151. State of Georgia House Bill No 472, https://www.legis.ga.gov/api/legislation/document/20172018/170675 (last visited December 24, 2020). "

152. Id.

153. State of Georgia House Bill No 219, https://www.legis.ga.gov/Legislation/20172018/170801.pdf (last visited December 24, 2020).

the dynamic driving task when the ADS is operating within its operational design domain"[154].

The bill further talks of driving license requirements at 40-5-21 para 11 and at 40-6-279 talks of actions to be taken by an autonomous vehicle[155] in case of an accident to remain compliant with the law. As per para 5.1:

"Automated driving system means the hardware and software that are collectively 16 capable of performing the entire dynamic driving task on a sustained basis, regardless of 17 whether it is limited to a specific operational design domain"[156].

2.6.4 *Louisiana*

In May 2018 vide HB 308, Louisiana allowed the operation of commercial vehicles in a platoon[157] after approval of the Department of Public Safety and Corrections and the Department of Transportation and Development[158]. As per Section 1 (95) of the bill Platoon or platooning denotes a collection of distinct motor vehicles, such as a truck, truck-tractor, trailer, semitrailer, or any combination of these, travelling together at close following distances using vehicle-to-vehicle communication equipment. Exceptions for following vehicles are (a) When travelling on a highway away from a commercial or residential area, a motor truck's driver is not allowed to follow another motor truck within 400 feet, however, this rule is not to be interpreted in a way that prevents one motor truck from passing and overtaking another, (b) Motor vehicles must be operated so that there is enough room between each one or combination of vehicles when they are driven in a motorcade or caravan on any road that isn't in a commercial or residential area, whether or not they are towing other vehicles[159]. This space must be left open to make it safe for any other vehicle to drive into and stay there. A funeral procession is exempt from this clause's restrictions.

[154.] Id.

[155.] Id.

[156.] Id.

[157.] Louisiana House Bill No 308 [online] available at https://custom.statenet.com/public/resources.cgi?id=ID%3Abill%3ALA2018000H308&ciq=ncsl&client_md=569a3280675b8f459612254d497b0650&mode=current_text (last visited December 26, 2020).

[158.] Id.

[159.] Id.

Further exceptions as per Section 1(95), if the Department of Public Safety and Corrections, the Office of State Police, and the Department of Transportation and Development have all authorised the platoon operator's operational plan, the platoon may be operated under this Section. Rules may be adopted by the Department of Public Safety and Corrections, the Office of State Police and the Department of Transportation and Development to carry out the obligations of this subsection. Platoon activities are not permitted on two-lane highways, and this Section's rules do not apply when a non-lead motor vehicle is involved[160].

2.6.5 *Maine*

In January 2018 the Governor of Maine passed executive order and created a Highly Automated Vehicles (HAV) Advisory Committee[161]. The advisory council would be diverse under the governor's plan, including at least 11 members, and include the commissioner of the Maine Department of Transportation, the secretary of state, the Bureau of Insurance, and the Office of Aging and Disability[162]. The role of this committee was to supervise the adoption of automated vehicles. In addition, the committee also was mandated to evaluate, create, and put into practice suggestions for initiatives that enhance these technologies. The order also mandated that no AVs must be tested without permits. Thereafter on 10 April 2018, the State of Maine vide Bill H.P. 1204 - L.D. 1724 established a Commission on AVs for allowing the testing, demonstration, and deployment of Automated Vehicles[163]. As per Section 2 of the bill, any hardware and software combination that can constantly complete the entire dynamic driving task is referred to as an automated driving system, regardless of whether it is restricted to a specific operational design domain.[164]. According to the norms and requirements defined in standard J3016, which the Society of Automotive Engineers adopted in September 2016, an automated

---

[160] Louisiana HB 308 [online] available at https://legis.la.gov (last visited September 22, 2021).

[161] Maine Executive Order 2018-001 [online] available at https://www.maine.gov/mdot/autonomous-vehicles/docs/2018-001_HAVExecOrder.pdf (last visited February 23, 2022).

[162] Id.

[163] State of Maine HP1204-LD1724 [online] available at https://mainelegislature.org/legis/bills/getPDF.asp?paper=HP1204&item=3&snum=128 (last visited December 26, 2020).

[164] Id.

driving system is particularly used to designate a level 3, level 4, or level 5 driving automation system.

Further as per Section 2 of the bill any vehicle or motor vehicle with a driving automation system is considered an autonomous vehicle[165]. A person or organisation that creates or sells AVs, as well as those that create or integrate automated driving technologies into motor vehicles that weren't designed to be AVs, are referred to as autonomous vehicle manufacturers. An autonomous vehicle manufacturer, higher education institution, fleet service provider, or supplier of automotive equipment or technology is referred to as an autonomous vehicle tester. A system in a motor vehicle that continuously executes all or a portion of the dynamic driving duty is referred to as a driving automation system. The commission vide the bill has been mandated to come up with a process to authorise testing of AVs, review applicable existing laws, recommend changes to laws, and develop a compliance framework[166]. This bill also permitted Transportation Commissioner to prohibit testing if there was a possibility of public safety risk.

2.6.6 Michigan

In December 2013 vide Senate Bill 663[167] and in March 2014 vide Senate bill 169, Michigan permitted testing of automated vehicles under laid down conditions[168]. The bill amongst other definitions also defined Automated Motor Vehicle and Manufacturer of Automated Technology. Under Sec 817, the Bill also provided immunity from civil liability to the manufacturer of automated technology for damages caused due to changes made by a third party to a motor vehicle/automated motor vehicle. In December 2016 vide Act 332 Michigan passed three more bills related to AVs. Section 665 (5) (Bill 996)[169] states that when activated, an automated driving system that is allowed to run without a driver must be treated as a driver to comply with applicable

---

[165]. Id.

[166]. Id.

[167]. State of Michigan enrolled Senate Bill no 663, https://www.legislature.mi.gov/documents/2013-2014/publicact/pdf/2013-PA-0251.pdf (last visited December 26, 2020).

[168]. State of Michigan enrolled Senate Bill no 169, https://www.legislature.mi.gov/documents/2013-2014/publicact/pdf/2013-PA-0231.pdf (last visited December 26, 2020).

[169]. State of Michigan enrolled Senate Bill no 995, https://www.legislature.mi.gov/documents/2015-2016/publicact/pdf/2016-PA-0332.pdf (last visited December 26, 2020).

motor vehicle laws. The 500-foot minimum separation standard does not apply to trucks in a platoon. As per Senate Bill 998 of March 2017[170], when automated vehicle repairs are carried out per manufacturer requirements, auto technicians and repair facilities are exempt from product liability.

2.6.7 Mississippi

Vide Bill no 1343 passed in April 2018 non-leading platoon vehicle was exempted from the too closely following law[171]. As per amended Section 63-3-103, Mississippi Code of 1972, the term "platoon" refers to a group of individual motor vehicles moving at electronically controlled speeds and closer following distances than would be practical and appropriate, absent such coordination. Further, as per amended Section 63-3-619, the driver of a motor vehicle shall not follow another motor vehicle closer than is reasonable and prudent, having regard to the other vehicle's speed, the volume of traffic, and the condition of the highway[172]. The driver of any motor truck or motor truck drawing another vehicle is not allowed to follow another motor truck or motor truck drawing another vehicle within three hundred (300) feet when travelling on a route outside of a business or residential area. These rules should not be interpreted to forbid passing or overtaking, nor should they apply to any lane that motor trucks are allowed to use exclusively.

The Motor Carrier Division was further directed to develop standards required for platoon plans[173]. The Code also prescribed at (3) (a) that:

"Subject to the provisions of paragraph (b) of this subsection, subsections (1) and (2) of this section shall not apply to the operator of a nonlead vehicle in a platoon, as defined in Section 63-3-103(k), as long as the platoon is operating on a limited access divided highway with more than one (1) lane in each direction and the platoon consists of not more than two (2) motor vehicles. (b) Only after an operator submits a plan for approval of general platoon operations to the Department of Transportation may a platoon be operated in this state. If that department approves the submission, it shall

---

[170.] Automated Vehicles driving and testing SB 995-998, https://www.legislature.mi.gov/documents/2015-2016/billanalysis/Senate/pdf/2015-SFA-0995-N.pdf (last visited December 26, 2020).
[171.] HB No 1343 [online] available at http://billstatus.ls.state.ms.us/documents/2018/html/HB/1300-1399/HB1343SG.htm (last visited December 26, 2020).
[172.] Id.
[173.] Id.

forward the plan to the Department of Public Safety for approval. Within thirty (30) days of filing, the plan must be examined and authorised or disapproved by the departments of transportation and public safety. If approved by both departments, the operator shall be allowed to operate the platoon five (5) working days after plan approval. The Department of Public Safety's Motor Carrier Division is responsible for creating the acceptable criteria needed for each section of the plan"[174] (Mississippi HB 1343).

2.6.8 Nebraska

In April 2018 vide Bill no LB 989, Nebraska allowed a driverless vehicle to operate on the roads of Nebraska[175]. This permission was granted subject to the vehicle having technological advancement to cause minimal risk in case of any malfunction. As per Section 1, an Automated driving system means all hardware and software, whether or not they are restricted to a particular operational design domain, that is capable of completing the complete dynamic driving task continuously. The vehicle was also mandated to be programmed to work in compliance with traffic laws. The bill[176] also mandated operators of these AVs to produce proof of proper insurance. There was also one more clause which specified that "AVs must stay at the place of accident in case any accident occurs". As per Section 1 of the bill:

"Automated-driving-system-equipped vehicle means a motor vehicle equipped with an automated driving system; Conventional human driver means a human person who manually exercises in-vehicle braking, accelerating, steering, and transmission gear selection input devices in order to operate a motor vehicle; Department means the Department of Motor Vehicles; Driverless-capable vehicle means a motor vehicle equipped with an automated driving system capable of performing all aspects of the dynamic driving task within its operational design domain, if any, including achieving a minimal risk condition, without any intervention or supervision by a conventional human driver"[177](Nebraska Bill No LB 989).

---

[174] Id.
[175] Nebraska Legislative Bill 989 [online] available at https://legiscan.com/NE/text/LB989/2017 (last visited December 26, 2020).
[176] Id.
[177] Id.

In 2019 one more bill was submitted for approval and that was LB 142. This bill mandated that automated vehicle makers are required to carry $5 million in liability insurance as well as $1 million in insurance for each vehicle involved in an accident[178]. This bill was strongly opposed and got suspended.

2.6.9 *Nevada*

On 17 June 2011 vide Section 8 in Bill no 511 Nevada authorised the operation of AVs[179]. The bill commanded the Department of Motor Vehicles to formulate rules for driving licenses, operation, and testing of these AVs. The bill also permitted the usage of cell phones by people in legally operating AVs. Vide Senate Bill 313 passed in June 2013[180] Nevada required proof of insurance from the drivers of autonomous vehicles[181]. As per Section 2, of the bill:

"Autonomous technology means technology which is installed on a motor vehicle, and which has the capability to drive the motor vehicle without the active control or monitoring of a human operator. The term does not include an active safety system or a system for driver assistance, including, without limitation, a system to provide electronic blind spot detection, crash avoidance, emergency braking, parking assistance, adaptive cruise control, lane keeping assistance, lane departure warning, or traffic jam and queuing assistance, unless any such system, alone or in combination with any other system, enables the vehicle on which the system is installed to be driven without the active control or monitoring of a human operator"[182] (Nevada SB 313).

Further as per Section 5 "unless the defect that caused the injury was present in the vehicle as originally manufactured, the manufacturer of a motor vehicle that has been converted by a third party into an autonomous vehicle

---

[178.] Nebraska senator floats bill to address driverless vehicle liability, Roads & Bridges, https://www.roadsbridges.com/nebraska-senator-floats-bill-address-driverless-vehicle-liability (last visited February 16, 2021).

[179.] Nevada AB511: 2011: 76th Legislature, Legiscan, https://legiscan.com/NV/text/AB511/2011 (last visited December 26, 2020). The bill in Section 8 (3) (b) defined Autonomous Vehicle as a "motor vehicle using artificial intelligence, sensors and GPS coordinates to drive itself without the active intervention of a human operator".

[180.] Nevada SB 313 [online] available at https://legiscan.com/NV/text/SB313/2013 (last visited December 26, 2020).

[181.] Id.

[182.] Id.

is not liable for damages to any person injured due to a defect caused by the conversion of the motor vehicle or by any equipment installed to facilitate the conversion"[183].

Section 4 of SB 313[184] required Autonomous vehicles to comply with all applicable laws, rules, and regulations before being permitted to register in the state. In June 2017 vide Assembly Bill 69 Nevada allowed plying of autonomous platoon trucks on state highways[185]. The bill specified that the "minimum distance rule does not apply to vehicles in a platoon". As per Section 2:

"Driver-assistive platooning technology means technology which enables two or more trucks or other motor vehicles to travel on a highway at electronically coordinated speeds in a unified manner at a following distance that is closer than would be reasonable and prudent without the use of the technology"[186] (Nevada SB 313).

Also, fines up to $2500 were prescribed in the said bill for violation of traffic rules by AVs. Carriers and Taxi operators were allowed to use AVs. The bill also defined Autonomous Vehicle Networking Company and laid now legal provisions concerning the operation of such companies[187].

2.6.10 *Oregon*

Vide Bill no 4059 passed in March 2018, Oregon[188] exempted the driver of a vehicle who was part of a connected automated braking system in a platoon from the too-close law. Section 40. ORS 811.485 was amended to read:

"811.485. (1) If a person does any of the following, they are guilty of following too closely: (c) Drives a motor vehicle when travelling upon a roadway outside of a business or residence district or upon a freeway within

---

[183.] Id.

[184.] State of Nevada Senator Bill No 313, https://www.leg.state.nv.us/Session/77th2013/Bills/SB/SB313_EN.pdf (last visited December 26, 2020). n

[185.] State of Nevada Assembly Bill No 69, https://www.leg.state.nv.us/Session/79th2017/Bills/AB/AB69_EN.pdf (last visited December 26, 2020).

[186.] Id.

[187.] Id. As per Section. 2.5. "Fully autonomous vehicle" means a vehicle equipped with an automated driving system which is designed to function at a level of driving automation of level 4 or 5 pursuant to SAE J3016.

[188.] Legislative Assembly of Oregon House Bill no 4059, https://legiscan.com/OR/text/HB4059/2018 (last visited December 28, 2020).

the corporate limits of a city in a motorcade or caravan, whether or not dragging another vehicle, without operating the vehicle to provide enough room between vehicles so that a vehicle can enter and occupy the space safely. (2) A funeral procession is not covered by this section. Vehicles taking part in a funeral procession must follow the lead vehicle as closely as is reasonable and safe, with the exception of the funeral lead vehicle"[189] (Oregon HB 4059).

The bill also established a task force to coordinate programs and policies on AVs. Additionally, the task committee was requested to research and offer recommendations on various aspects of AVs including licensing, registration, reporting of accidents, insurance, and liability[190]. As per (3)(a) of the bill a driver of a vehicle that is a part of a connected autonomous braking system is exempt from the provisions of this section. In this subsection, the term connected automated braking system refers to a system that electronically synchronises one or more following vehicles' brakes with the lead vehicle's brakes using vehicle-to-vehicle communication[191]. The task force was also expected to provide suggestions on infrastructure designs for facilitating AVs to run on the roads of Oregon.

2.6.11 *Tennessee*

Vide Senate Bill 1561 dated 01 July 2016 Tennessee amended its laws relating to automotive vehicles specially to add more details concerning AVs[192]. As per section 1 (b) (1):

"Autonomous technology means technology installed on a motor vehicle that has the capability to drive the vehicle on which the technology is installed in high or full automation mode, without any supervision by a human operator, with specific driving mode performance by the automated driving system of all aspects of the dynamic driving task that can be managed by a human driver, including the ability to automatically bring the motor vehicle into a minimal risk condition in the event of a critical vehicle or system failure or other emergency event"[193] (Tennessee SB 1561).

---

[189] Id.

[190] Id.

[191] Id.

[192] State of Tennessee Senate Bill No 1561, https://publications.tnsosfiles.com/acts/109/pub/pc0927.pdf (last visited December 28, 2020).

[193] Id.

In April 2017 Tennessee permitted the operation of vehicles in a platoon with requisite automated safeguards following the operator's notification to the Department of Transportation and Safety[194]. In June 2017 vide SB 151 Tennessee created the Automated Vehicles Act. The Act[195] was made applicable only to AVs in high or full automation mode. Vide Amendment #2 of this bill a manufacturer owner of an Automated Driving System (ADS) operated vehicle shall maintain primary automobile liability insurance with minimum limits of $5 million for bodily injury and property damage[196]. The Act had provisions about seat belts, kid safety seats, abandoned vehicles, and AV crash reporting. The Act permitted AVs without a driver under certain conditions. For liability[197], the Act specified Automated Vehicle as a driver in high or full automation mode.

2.6.12 *Texas*

Vide HB 1791 on 01 September 2017, Texas[198] allowing networked, electronically coordinated braking to be used to keep vehicles at the right distance from one another[199]. Section 545.062, of the Transportation Code, was changed by introducing Subsection (d), which allowed a driver of a vehicle utilising a connected brake system to receive help from the system to maintain the necessary assured clear distance or amount of space while following another vehicle using that system[200].

---

[194] Vehicle Platooning [online] available at https://www.tn.gov/tdot/transportation-freight-and-logistics-home/vehicle-platooning.html (last visited December 28, 2020). Platooning is the linking of two or more vehicles in a convoy using wireless communications and sensor technology.

[195] Tennessee SB151, Track Bill, https://trackbill.com/bill/tennessee-senate-bill-151-motor-vehicles-as-enacted-enacts-the-automated-vehicles-act-and-other-requirements-related-to-the-operation-of-autonomous-vehicles-on-the-public-roads-of-this-state-amends-tca-title-5-title-6-title-7-title-39-title-40-title-54-title-55-title-56-title-65-and-title-67/1375094/#/details=true (last visited December 28, 2020).

[196] Id.

[197] Id. As regards nonmanufacturer owner the requirement is "to maintain primary automobile liability insurance providing at least: $50,000 for death or bodily injury, per person; $100,000 for death or bodily injury, per incident; and $30,000 for property damage".

[198] State of Texas HB 1791, https://capitol.texas.gov/tlodocs/85R/billtext/pdf/HB01791F.pdf#navpanes=0 (last visited December 28, 2020).

[199] Id. In this subsection (d) of Section 545.062, "connected braking system" means a system by which the braking of one vehicle is electronically coordinated with the braking system of a following vehicle.

[200] Id.

Vide HB 2205 it was also specified[201] that once an AV is engaged then the owner of that AV shall be the operator of the AV. As per Sec.545.453:

"Operator of Automated Vehicle means (a)When an automated driving system installed on a motor vehicle is engaged: (1) the owner of the automated driving system is considered the operator of the automated motor vehicle solely for the purpose of assessing compliance with applicable traffic or motor vehicle laws, regardless of whether the person is physically present in the vehicle while the vehicle is operating; and (2) the automated driving system is considered to be licensed to operate the vehicle. (b)Regardless of other laws, if a vehicle's automated driving system is activated, a licensed human operator is not necessary to operate it"[202] (Texas HB 2205).

Bill 2205 also specified that the system shall be considered as licensed to operate the vehicle[203] and defined ADS in Section 545.451 as installed hardware and software on a motor vehicle that when activated can perform tasks without any assistance from or oversight from a human operator[204].

2.6.13   Utah

This state started in the AV space very slowly and cautiously. In May 2015 Utah authorised the Department of Transports to carry out tests on connected vehicle technology. Subsequently, in May 2016, the state mandated a study on Automated Vehicles to evaluate the American Association of Motor Vehicle Administrators (AAMVA) and the National Highway Traffic Safety Administration (NHTSA) standards to figure out mandatory safety features and make recommendations. In May 2018 the state allowed vehicles with Connected Platooning Systems in place to ply on the roads of Utah.

On 29 March 2019, the State of Utah passed a bill, HB 101, to allow full AVs to operate on Utah roads[205]. The measure lays out rules for self-driving cars as well as protocols for handling collisions. Since the vehicles will

---

201. State of Texas SB 2205, https://capitol.texas.gov/tlodocs/85R/billtext/pdf/SB02205F.pdf#navpanes=0 (last visited December 28, 2020).

202. Id.

203. Id.

204. Id.

205. Emily Ashcraft, Legislature approves bill allowing driverless cars on Utah roads Deseret News (2019), https://www.deseret.com/2019/3/1/20667185/legislature-approves-bill-allowing-driverless-cars-on-utah-roads (last visited December 28, 2020).

initially be used for testing, the bill also exempts them from licensing. This bill has language that allows driverless automated vehicles plied by companies like Uber or Lyft[206]. As per the bill a motor vehicle with a level four or level five ADS is permitted to travel autonomously on a highway in the state as long as the ADS, or Automatic Driving System (as defined in the statute HB 101), complies with all applicable state and federal laws regarding traffic and motor vehicles[207]. Bill HB 101 talks of vehicle data being private data of the vehicle owner[208]. According to the statute, the owner of the car is the exclusive owner of the private data that the vehicle collects. However, as long as the owner, passenger, or human driver's identity is kept a secret, data sharing is permitted to enhance motor vehicle safety, security, or traffic management. Utah has finally caught up to California, Arizona, and Pennsylvania, three states whose legislation permits the testing of automated vehicles on public highways[209].

2.6.14 Vermont

As early as May 2017 Vermont required the Department of Transports to work with experts on AVs to come up with proposed recommendations and legislations for plying of AVs in the State. In June 2019 Vermont came up with the Automated Vehicles Testing Act[210]. As per definitions in § 4202 "the term automated driving system refers to technology and software that, when used together, are capable of performing the complete dynamic driving task continuously and, if necessary, reaching a low-risk condition without the aid or supervision of a conventional human driver"[211].

---

[206] Utah HB0101: 2019: General Session, Legiscan, https://legiscan.com/UT/text/HB0101/2019 (last visited December 28, 2020).

[207] Id.

[208] Betsy Foresman, Utah to join states allowing fully AVs on public roads StateScoop (2019), https://statescoop.com/utah-to-join-states-allowing-fully-autonomous-vehicles-on-public-roads/ (last visited December 28, 2020).

[209] Self-driving cars get the greenlight under new utah law government technology state & local articles - e.republic, https://www.govtech.com/policy/Self-Driving-Cars-Get-the-Greenlight-Under-New-Utah-Law.html (last visited December 28, 2020).

[210] Vermont Laws, State House Dome, https://legislature.vermont.gov/statutes/fullchapter/23/041 (last visited December 28, 2020).

[211] Id.

This Act was instrumental in bringing about a process for testing automated vehicles[212]. According to the Act, to operate an automated vehicle, testing authorization must be granted by the Traffic Committee, which is comprised of the Secretary of Transportation, the Commissioner of Motor Vehicles, and the Commissioner of Public Safety[213]. The Act also elaborates that ADS enables HAVs to operate with minimal human assistance. These ADS vehicles analyse and recognise their environment using cameras, radar, lidar (image sensing), GPS, and computer vision.[214]. The Act recognises that there are five degrees of automation, and level 3, level 4, or level 5 automated vehicles cannot be tested on state or town highways without a permit from the Vermont Traffic Committee (highly automated).

2.6.15   Other States in the USA

In April 2016 the Joint Legislative Committee ("Committee") was established in Alabama to carry out a study on AVs.[215] The Committee found lots of legal issues and one of them concerning a blind person who cannot operate a conventional vehicle but would be able to use a self-driving car. The question was can blind persons be then given a driving license. That's just one aspect. State Sen. Gerald Allen (R-Tuscaloosa) sponsored house bill HB 47 on 05 March 2019 seeking approval to allow AVs operated by an automated driving system (ADS)[216]. As per the bill, the ADS must operate the vehicle safely even if there is a partial systems failure. The Committee also observed that for efficient and safe working of AVs "some amount of infrastructure[217] like satellites, road striping, interactive signs and traffic signals are mandatory to

---

[212] Id. As per the Act, "Automated vehicle" means a motor vehicle that is equipped with an automated driving system".

[213] Id.

[214] Automated Vehicle Testing in Vermont | Agency of Transportation, https://vtrans.vermont.gov/planning/av-testing#:~:text=Vermont's%20Automated%20Vehicle%20Testing%20Act,on%20state%20and%20town%20highways, (Last visited December 28, 2020).

[215] William Thornton | wthornton@al.com, Alabama Legislature preparing for self-driving cars AL (2019), https://www.al.com/news/anniston-gadsden/2019/03/how-the-alabama-legislature-is-preparing-for-self-driving-cars.html (last visited December 17, 2020).

[216] How the Alabama Legislature is preparing for Self-Driving Cars? [online] available at https://www.al.com/news/anniston-gadsden/2019/03/how-the-alabama-legislature-is-preparing-for-self-driving-cars.html (last visited February 23, 2022). Allen said his bill, HB47 introduced March 5, still has several issues to work out. But it illustrates some of the challenges lawmakers face.

[217] Id.

help vehicles figure out when and where to adjust the vehicle speed and how to change lanes or allow other vehicles to change lanes". Allen's bill notes that a traditional human driver does not have to be in the car at all times while using an autonomous driving system. However, the presence of a driver would enable a fully autonomous driving system that can continue to function securely even if some of its systems fail partially. Additionally, vehicles must be equipped with recording equipment and have insurance. Additionally, self-driving legislation must apply to the trucking sector as well as to the general driving public[218]. Accordingly, the law needs to be tweaked.

The State of Arkansas in their 91st General Assembly in 2017 vide House Bill 1754 approved the testing of AVs and allowed for self-driving truck platooning systems[219]. As per the bill, driver-assistive truck platooning systems use sensor arrays, wireless connection, vehicle controls, and specialised software to coordinate the acceleration and braking of two (2) or more vehicles. A person requiring plying such truck platooning systems must seek permission for platoon operations from the State Highway Commission[220]. The joining of two or more trucks in a convoy utilising connectivity technology and autonomous driving assistance systems is known as truck platooning. These trucks automatically keep a predetermined, close distance from one another while they are connected during portions of the journey[221]. Driver intervention is minimal to nonexistent because the truck at the front of the platoon functions as the leader and the trucks following it respond to changes in its movement. Drivers will always be in charge in the first instance, giving them the option to quit the platoon and drive alone.

Colorado State passed Senate Bill 17-213 to authorize ADS to control motor vehicles in the state of Colorado. As per the Senate, "in 2016, more than 600 people died in Colorado due to road accidents caused by human error". Since ADS can reduce accidents caused due to human errors the state was keen on testing and deploying automated vehicles in Colorado[222]. The bill under

---

[218.] Id.

[219.] State Of Arkansas House Bill 1754, https://www.arkleg.state.ar.us/Acts/Document?type=pdf&act=797&ddBienniumSession=2017%2F2017R (last visited December 20, 2020).

[220.] What is truck platooning (2017), https://www.acea.be/uploads/publications/Platooning_roadmap.pdf (last visited December 22, 2020).

[221.] Id.

[222.] State of Colorado Senate Bill 17-213 [online] available at http://leg.colorado.gov/sites/default/files/2017a_213_signed.pdf (last visited December 24, 2020)."

42-1-102 defines ADS as hardware and software that can collectively perform all aspects of driving for Levels 4 and 5 automations in SAE International Standards without the need for human participation[223]. As per the Senate "such testing and deployment would also help bring about a reputation for Colorado as a hub for advanced technologies". The bill also prescribed that the Colorado DOT shall provide yearly test reports to the Transportation Review Committee. The bill also mentions that the liability for any accident due to a driverless automated vehicle shall be decided according to the relevant State, Federal, and Common Law.

Vide Substitute Senate Bill No. 260, Public Act No. 17-69, approved on 27 June 2017, Connecticut passed an act concerning AVs[224]. The act defines a fully autonomous vehicle as a motor vehicle having level four or level five ADS. As per Section 1 (a) a vehicle that has an automated driving system and is fully self-driving is referred to as a fully autonomous vehicle and classified as level four or level five by SAE standards[225]. The act requires the vehicle operator to be seated in the driver's seat and be insured for at least $5 million. The act has also established a task force to carry out a study on AVs[226] and evaluate NHTSA's standards for AVs and laws in other states for making appropriate recommendations. As per the act, "Automated driving system, means the hardware and software that, taken together, are capable of consistently completing the complete dynamic driving task, regardless of whether the automated driving system is only capable of operating inside a certain operational design domain and dynamic driving task indicates the urgent tactical and operational tasks necessary to drive a car on the highway, eliminating the strategic duties like planning trips and choosing locations and waypoints"[227] (Connecticut Public Act 17-69).

Vide DC Act 19-643 passed on 23 January 2013, the Department of Motor Vehicles was required to establish safe operating protocols for facilitating AVs to ply in the District of Columbia[228]. As per Section 4 of the Act, a car's

---

[223] Id.
[224] An act concerning AVs., https://www.cga.ct.gov/2017/Act/pa/2017PA-00069-R00SB-00260-PA.htm (last visited December 24, 2020).
[225] Id.
[226] Id.
[227] Id.
[228] DC Act 19-643 [online] available at https://lims.dccouncil.us/downloads/LIMS/26687/Signed_Act/B19-0931-SignedAct.pdf (last visited February 23, 2022).

original manufacturer is not responsible for any accidents brought on by a vehicle that a third-party vendor converts to an autonomous vehicle. Further, as per Section 3, a human driver must be there and prepared to take over in an autonomous vehicle, if necessary.[229]. The act also defined an automated vehicle (AV) as a vehicle that can operate without a driver and is capable of navigating through roads and reading traffic signals[230]. In July 2018, the DC Senate granted a year to the DOT to study and evaluate AVs and make the study publicly available. Vide Public Act 100-0352 in August 2017, Illinois laid down detailed provisions which prevent local authorities from prohibiting AVs[231]. A vehicle that has an automated driving system installed, per sub-section e-5, is capable of performing the entire dynamic driving task continuously and consists of both hardware and software, regardless of whether it is restricted to a certain operational domain[232]. This clause (e-5) imposes restrictions on the concurrent exercise of powers and duties by home rule entities per Illinois' Constitution's Article VII, Section 6, subparagraph I.

Since safety was a prime concern, Indiana was a bit slow in moving a bill to permit automated vehicles to get tested and ply in the State of Indiana[233]. The Indiana House of Representatives unanimously passed bill no HB 1341[234] to work on a framework for operating AVs in the state. The legislation mandated that anybody driving an autonomous vehicle in Indiana register with the state's Bureau of Motor Cars (BMV) and that automated vehicles adhere to all current local, state, and federal laws. Additionally, operators would need to file $5 million in financial responsibility with the BMV for each car being tested. The 300-foot limitations on proximity in case of vehicle platooning were also exempted by the bill[235].

In March 2018, vide SB 116, Kentucky permitted platoon plying of commercial vehicles after giving notice to the Department of Vehicle Regulation (DVR) and the Kentucky State Police. One of the mandatory ask

---

[229.] Id.

[230.] Id.

[231.] State of Illinois Public Act 100-0352, https://www.ilga.gov/legislation/publicacts/100/Pdf/100-0352.pdf (last visited December 24, 2020).

[232.] Id.

[233.] Colin Wood, Slow and steady, Indiana readies autonomous vehicle framework StateScoop (2018), https://statescoop.com/slow-and-steady-indiana-readies-autonomous-vehicle-framework/ (last visited December 24, 2020).

[234.] Id.

[235.] Id.

in this bill is the requirement to have a driver with a valid commercial driver's license in each vehicle that is part of a platoon[236]. According to Section 2 (39) of the bill, a platoon is defined as a group of two (2) separate commercial motor vehicles travelling together at electronically coordinated speeds and following distances that are closer than would ordinarily be allowed under subsection (8)(b) of Section 3 of this Act[237]. On 23 January 2017 New York passed a bill[238] permitting the testing and use of AVs in demonstrations so that the commissioner of motor vehicles can better understand how such technology may affect safety, traffic regulation, traffic enforcement, and emergency services. The dynamic driving task is defined as all of the real-time operational and tactical functions necessary to operate a vehicle in on-road traffic, excluding strategic functions like trip planning and destination and waypoint selection. According to Section 1(b), autonomous vehicle technology is defined as hardware and software that collectively are capable of performing part or all of the dynamic driving task on a sustained basis.[239]. Bill no A 9508 dated 18 January 2018[240] lays down the requirement of specifying the procedure for the first respondents to interact during an emergency with AVs.

On 21 July 2017 vide Bill no 469[241] North Carolina passed a law governing the use of fully automated vehicles. As per Article 18 Section 20-400 of the Act:

"Fully Autonomous Vehicle means a motor vehicle equipped with an ADS that will not at any time require an occupant to perform any portion of the dynamic driving task when the ADS is engaged. If equipment that allows an occupant to perform any portion of the dynamic driving task is installed, it must be stowed or made unusable in such a manner that an occupant cannot

---

[236]. Kentucky SB 116 [online] available at https://apps.legislature.ky.gov/law/acts/18RS/documents/0033.pdf (last visited February 23, 2022).

[237]. Id.

[238]. State of New York S 2005-C, A 2005-C, https://nyassembly.gov/2017budget/budget_bills/A3005C.pdf (last visited December 26, 2020).

[239]. Id.

[240]. State of New York Bill no A09508C [online] available at https://nyassembly.gov/leg/?default_fld=&leg_video=&bn=A09508&term=2017&Summary=Y&Text=Y (last visited December 26, 2020). AVs which are getting tested to follow guidelines from the police superintendent.

[241]. General Assembly of North Carolina - House Bill No 469, https://legiscan.com/NC/text/H469/2017 (last visited December 28, 2020).

assume control of the vehicle when the ADS is engaged"[242] (North Carolina HB No 469).

As per Article 18, Section 20-401, an autonomous vehicle's driver is not obliged to have a driver's license. An adult must ride in an autonomous vehicle with a passenger under the age of twelve". The statute requires the creation of a committee for fully autonomous vehicles, and the subsequent too-close law was changed to permit vehicle platooning[243].

Vide Bill no 1065 dated 06 January 2015, North Dakota[244] commenced a study on AVs. The study was focused on finding the reduction in fatal accidents and crashes that AVs could bring about. The study also focused on reducing congestion and improving fuel economy through AVs. Vide Bill no 1202 dated 03 January 2017, North Dakota[245] mandated the Department of Transports to gather data on the usage of AVs on highways and analyse them. The bill also mandated a review of existing regulations on data ownership/use, inspection, insurance, and registration for motor vehicles and make recommendations on how to adopt them for AVs. Although Pennsylvania does not have specific laws for AVs nor has permitted their plying on the state roads, on 20 July 2016 vide Bill no SB 1267 the state-sanctioned a usage of up to $40 million[246] to develop technology connected with AVs.

Vide Bill no HB 3289 South Carolina on 19 May 2017 declared that "the law regarding minimum following distance[247] does not apply to vehicles travelling in a platoon". As per Section 56-5-1930:

"(A) Having appropriate consideration for the speed of such vehicles, the volume of traffic on the road, and the state of the roadway, the driver of a motor vehicle shall not follow another vehicle more closely than is reasonable and sensible. (B) When conditions allow, the driver of any truck

---

[242.] Id.

[243.] Id. As per Article 18, Section 20-401, Driver's License is not required for the operator of a fully Autonomous Vehicle. An adult to be present in an Autonomous Vehicle in which a child under twelve years was travelling.

[244.] Legislative Assembly of North Dakota House Bill No 1065, https://www.legis.nd.gov/assembly/64-2015/documents/15-0167-03000.pdf (last visited December 28, 2020).

[245.] Legislative Assembly of North Dakota House Bill no 1202, https://www.legis.nd.gov/assembly/65-2017/documents/17-0711-04000.pdf (last visited December 28, 2020).

[246.] Legislative Data Processing Center, 2016 ACT 101, The official website for the pennsylvania general assembly. https://www.legis.state.pa.us/cfdocs/legis/li/uconsCheck.cfm?yr=2016&sessInd=0&act=101# (last visited December 28, 2020).

[247.] 2017-2018 BILL 3289: Safe following distance, https://www.scstatehouse.gov/sess122_2017-2018/bills/3289.htm (last visited December 28, 2020).

or motor vehicle drawing another vehicle that is following another truck or motor vehicle drawing another vehicle on a roadway outside of a business or residential district must leave enough space for the overtaking vehicle to enter and occupy the space without danger. However, this rule does not prevent a truck or motor vehicle drawing another vehicle from overtaking and passing any vehicle"[248] (South Carolina Bill No HB 3289).

This is in sync with a similar law in place in many states in the USA. An announcement in 2015 by then-Gov. Terry McAuliffe declared that Virginia was open for autonomous cars. The Virginia Government collaborated with Virginia Tech's Transportation Institute (VTTI) to launch the Virginia Automated Corridors initiative. This initiative was purely aimed to allow government agencies and private companies to test AV technology in Virginia. The testing will also be permitted to be done on interstate and artery roads of North Virginia (NoVA)[249]. The Virginia General Assembly passed a law in April 2016 allowing an automated vehicle's safety driver to watch a pre-installed video display from the front seat while the AV is moving[250]. Additionally, according to this law, government license plates are not required for automobiles utilised by universities for technological research.

Executive Order 17-02 signed in June 2017 tasked relevant agencies with promoting secure testing and use of AVs in Washington. The Executive Order included two main provisions regarding the formulation of a workgroup and conducting of pilot programs for AV testing[251]. The work group would consist of representatives from relevant agencies such as the Departments of Transportation, Commerce, and Licensing; the Transportation Commission; and the Office of Regulatory Innovation and Assistance, for the safe development of automated technology in vehicles and public roads, and to examine emerging automated transportation technology across various modes of transportation. Pilot programs were to enable to carry out safe testing and

---

[248.] Id.

[249.] Self-driving vehicles have officially arrived in Virginia, NORTHERN VIRGINIA MAGAZINE (2019), https://northernvirginiamag.com/culture/news/2019/07/05/the-future-of-self-driving-vehicles-has-officially-arrived-in-northern-virginia/#:~:text=Perhaps%20even%20more%20importantly%2C%20 a,for%20business%E2%80%9D%20for%20autonomous%20cars. (Last visited December 28, 2020).

[250.] Josh Mandell @joshuamandell et al., Virginia still a blank slate for self-driving car laws Charlottesville Tomorrow â€¢ Informed citizens create better communities, https://www.cvilletomorrow.org/articles/virginia-blank-slate-for-self-driving-car-laws (last visited February 20, 2021).

[251.] Bill Covington, Legislating AVs In Washington (2018), https://wstc.wa.gov/wp-content/uploads/2019/11/2018-0717-BP7-UWFullReportAVLawScan.pdf (last visited December 30, 2020).

operation of AVs. In May 2017 the Wisconsin Governor passed an executive order on the testing and deployment of AVs by creating a Governor's Steering Committee. This committee was tasked with coordinating with various agencies for registration, licensing, insurance, traffic laws, and criteria for equipment. The committee was also tasked with reviewing laws that impede testing/deployment. Vide Act 294 of 2017 Wisconsin provided an exception for platoons from the 500 feet minimum distance law for trucks weighing more than ten thousand pounds[252].

2.6.16   *States with Executive Orders in the USA*

In August 2015 the Arizona Governor issued an executive order directing various agencies to help in the testing and operation of AVs on public roads[253]. The Governor also ordered that pilot programs may be permitted at select universities. Thereafter in March 2018 the Governor again issued an updated order mandating Arizona to step up towards the latest technological developments in the field of automated vehicles and ensure that all automated vehicles in the state follow the Federal and State guidelines[254]. Between September 2017 to January 2018 Governors of Delaware, Hawaii and Idaho issued executive orders concerning AVs and their testing, recommendations. Idaho order also tasked the newly created Autonomous and Connected Vehicle Testing and Deployment Committee for facilitating the testing and use of AVs and work on registration, licensing, insurance, and traffic regulations[255]. In October 2016 the Governor of Massachusetts issued an executive order to promote the testing and deployment of AVs. Similar orders were passed in March 2018 in Minnesota and in January 2018 in Ohio. All of these directives were made to research, evaluate, and get ready for the introduction

---

[252] Report of the Governor's Steering Committee on Autonomous and Connected Vehicle Testing and Deployment [online] available at https://wisconsindot.gov/Documents/about-wisdot/who-we-are/comm-couns/av-final-report-062918.pdf (last visited December 30, 2020).

[253] Executive Order 2015-09 [online] available at https://azgovernor.gov/file/2660/download?token=nLkPLRi1 (last visited December 30, 2020).

[254] Arizona says human-free driving is a-okay https://www.theverge.com/2018/3/2/17071284/arizona-self-driving-car-governor-executive-order (last visited December 30, 2020). The article explains that as per the executive order, fully driverless cars without anyone behind the wheel are allowed to operate on public roads. The only caveat is that the vehicles follow all existing traffic laws and rules for cars and drivers.

[255] Autonomous Vehicle Federal Regulation [online] available at https://www.foley.com/en/insights/publications/2019/01/autonomous-vehicle-federal-regulation (last visited December 30, 2020)

of automated vehicles, which would reduce accidents[256]. Washington and Wisconsin also have executive orders issued by Governors relating to AVs[257].

2.7 Regulations of Self-Driving Cars or AVs in Major EU Nations

The creation of a uniform EU regulation for AVs will take time and one can expect that to be a gradual and slow process. Many countries in the EU are slowly and steadily working on regulations/laws for AVs. The AV manufacturers are putting countries under pressure to come up with laws for AV testing and operation on roads. Amendment of existing motor vehicle rules seems to be the common track adopted by most nations.

2.7.1 Austria

In March 2018, Austria collaborated with Slovenia and Hungary to come up with a mutual agreement for developing and testing new vehicle technologies including AVs. As per the agreement the testing was agreed to be conducted in a new Austrian-Hungarian-Slovenian driverless region[258]. Under its first Automated-Connected-Mobile Action Plan, Austria has come up with a legal regime for testing AVs and doing research on AVs with an investment of around 25 million euros[259] and with over 300 AV professionals. Austria has created a regulatory framework for the first Automated-Connected-Mobile Action Plan (2016-2018), as well as test environments and many research initiatives.

As per AV laws in Austria[260], testing is allowed for AV minibuses up to a max speed of 20 kilometres per hour The vehicle must have run a minimum of 1000 kilometres before testing and must have a safety driver to take control when required. Autopilot with an autonomous change of lanes is to be

---

[256] Id 75.

[257] Id 75.

[258] 2019 Autonomous Vehicle Readiness Index [online] available at https://assets.kpmg/content/dam/kpmg/be/pdf/2019/02/ADV-AutonomousVehicle2019-brochure-uk-LR.pdf (last visited March 2, 2021).

[259] Id.

[260] Austria paves the legal way for (partly) AVs [online] available at https://www.schoenherr.eu/content/austria-paves-the-legal-way-for-partly-autonomous-vehicles/ (last visited March 2, 2021).

engaged only on highways/freeways. Fully automated military vehicles which have run a minimum of 300 kilometres can be tested on public roads with a safety driver to take control when required. All AV testing[261] must have adequate Insurance coverage. According to the Ministry of Traffic, Innovation, and Technology's revision to the autonomous driving law (Automatisiertes Fahren Verordnung - AutomatFahrV), automated parking of automobiles is extensively permitted. Approval for testing is required from Federal Minister for Transport, Innovation, and Technology. Lastly, vehicle operators in the street where testing is to be carried out must be forewarned and adequately informed about the test.

In Austria, the country follows the general liability rules for any damage caused by an AV or semi-AV. The general Austrian law clearly distinguishes between liability for damages due to negligent behaviour and strict (product) liability of the car operator or the car manufacturer according to the product liability law[262]. As per Austrian law, the car operator would be responsible for any harm the vehicle may cause unless the car was used without the operator's consent. If any driver violates any rules the driver will be liable[263] for all damages. The operator as per Austrian law is liable both jointly and severally for failure to apply due care in preventing misuse of the car without consent (e.g., a mother does not lock her car and her child secretly drives away and hits a pedestrian). As regards damages to third parties or pedestrians normally the insurance will cover it. However, if in the future the law does not require driver attention in full AV there would be no liability on the driver per se.

Austrian law also provides for strict liability of the operator of any car irrespective of the fact that the car is autonomous or not. As per Austrian law, the car operator would be responsible for any harm caused by the car, unless the car was used without the operator's consent. The operator as per Austrian law is liable both jointly and severally for failure to apply due care in preventing misuse of the car without consent (e.g., a mother does not lock her car and her child secretly drives away and hits a pedestrian).

As per Austrian laws product liability also applies to software that's part of a physical product like an AV. Any bug or error in the software which

---

[261.] Id.

[262.] AVs Law and Regulations in Austria [online] available at https://cms.law/en/int/expert-guides/cms-expert-guide-to-autonomous-vehicles-avs/austria (last visited March 2, 2021).

[263.] Id.

results in an accident by an AV shall lead to product liability of the car manufacturer[264]. As regards AI software sometimes errors occur basis what the sensors decode. Things are still not very clear on how these situations will be handled especially from a product liability perspective. Imposing product liability on a manufacturer for this may not be the right approach, but the fact is every AV manufacturer carries this risk. Warranty claims can only be made in Austria against the seller and not against the manufacturer of the AV[265]. However, as mentioned above car manufacturers will be subject to strict liability under the applicable product liability laws.

2.7.2 *France*

The French government plans to highly deploy AVs on their roads between 2020 and 2022. Emmanuel Macron, the French President with a technological bent of mind tasked a very senior individual, Anne-Marie Idrac, with developing a national AV strategy[266], which included coming up with new laws and regulations for R&D or AVs, pilot projects and cyber security/privacy aspects of using AVs. All Level 3 and Level 4 passenger vehicles, autonomous mass transit like robotaxis, and automated delivery vehicles are projected to be permitted in France if the new laws are approved. The French government has allocated roughly around 40 million euros ($46 million) for subsidizing new AV-related projects[267]. Forty million euros have been made available by the government to assist in funding new projects.

One of the leading robotaxis developers in France is Nayva, a Lyon-based company, which operates a lot of self-driving shuttles across the world through pilot projects. Nayva has recently announced its decision to start an assembly plant in Michigan. In addition, Nayva also has small driverless shuttle buses[268]. These buses have been built by Nayva with the financial support of Valeo, a French supplier. Since 2017, Nayva AV buses have been plying on the roads

---

[264]. Id.

[265]. Id.

[266]. 2019 Autonomous Vehicle Readiness Index Supra note 277, at 71.

[267]. France pushes for 'highly automated' vehicles by 2022 Automotive News Europe, https://europe. autonews.com/article/20180808/ANE/180809840/France-pushes-for-highly-automated-vehicles-by-2022 (last visited March 2, 2021).

[268]. Id. More than 50 autonomous-vehicle test projects have taken place in France since 2014, including robotaxis, buses and private vehicles.

outside Paris in the area popularly known as La Defense which is the business zone. Since 2014, France has hosted more than 50 autonomous vehicle test programs, including those involving robotaxis, buses, and private automobiles.

As regards software for AVs, the PSA group has partnered with NuTonomy to create AV software. Plying in Singapore's business centre are Peugeot 3008 crossovers with NuTonomy's software. An offshoot of the Massachusetts Institute of Technology (MIT), Aptiv currently owns NuTonomy. The merger between Peugeot S.A. (Groupe PSA) and Fiat Chrysler Automobiles N.V. (FCA) which was announced in October 2019 finally became effective on 16 January 2021[269]. The French company Renault is also planning to partner with Nissan and put robotaxis on road by 2022[270]. As of date, France plans to permit Level 4 driverless AVs on its roads. Presently French laws only permit certain companies to test driverless AVs and this restriction is basically to ensure minimum risk to the public.

2.7.3 Germany

As mentioned at the beginning of this Chapter Germany being a pioneer nation for automobiles has a clear-cut established framework concerning AVs despite the absence of any adopted legislation at the EU level. The German legislation[271] is unique and is a role model for other EU countries to follow. As of date in Germany AV testing laws are handled by city regulatory authorities. Permission has been granted by a few authorities to pilot fleets on private property. AV regulation is being set by Germany, but AV deployment is another matter. Only fixed-route AVs are currently being tested in Germany, which is a good fit for a region that is primarily concerned with fixed-route AV use cases. The present German government has huge plans to create appropriate infrastructure which would be capable of handling Level 5 fully AVs[272]. The long-term potential for AVs in Germany and Europe is enormous,

---

[269] Merger of FCA and Groupe PSA https://www.fcagroup.com/en-US/media_center/fca_press_release/FiatDocuments/2021/January/The_merger_of_FCA_and_Groupe_PSA_has_been_completed.pdf

[270] France to amend legislation for autonomous vehicle trials France to amend legislation for autonomous vehicle trials | Autovista Group, https://autovistagroup.com/news-and-insights/france-amend-legislation-autonomous-vehicle-trials (last visited March 2, 2021)

[271] Germany first to pass AV Regulations [online] available at https://www.eetasia.com/germany-first-to-pass-av-regulations/ (last visited March 21, 2021).

[272] Id.

but they must contend with world-class mass transit systems. To improve service and save costs, many mass transit operators are ready to test, learn about, and implement fixed-route AVs. Innovative services that are used to reduce emissions from BEV-based AV usage will be developed through fixed-route use cases as well. The German government also has plans to extend the AV testing on the autobahn after the Bavarian A9 highway. In this area, the German government is also experimenting with communication between vehicles through 5G mobile networks.

Mercedes (Daimler), BMW, Audi, and Volkswagen are the leading German motor vehicle producers. These companies have come up with their AVs and have started testing them on open roads. Recently German lawyers also focussed on the issues arising out of the appearance and direction of AVs. These lawyers attempted to document a solution for the various questions on the appearance of AVs and the lack of existing laws.

In November 2015, the German government published a strategy[273] for modification of existing laws on road traffic to facilitate plying of AVs on German roads and thereafter proposed a draft to the Bundestag. The Bundestag is the German federal parliament which is elected by the people of Germany at the federal level. Per the proposal in the draft, the German Road Traffic Act, also known as the StVG or Straßenverkehrsgesetz, was updated in 2017. This amendment opened the path for the introduction of AVs[274] to road traffic in Germany. Traffic is predictable thanks to real-time data exchange between vehicles and infrastructure, which also helps to prevent accidents. By networking with the environment, vehicles become fully digitized mobility, information and communication platforms.

Article §1a of the StVG describes the fundamental difference between normal automated vehicles and fully automated vehicles as per German Laws[275]. The said article also states that a fully AV can be driven on German roads only if the fully AV function in the vehicle is used for what it is intended to serve and not for anything else. It also defines the framework for AVs as has been conceptualised within German laws. The basic premise under the StVG

[273.] Strategie automatisiertes und vernetztes Fahren BMVI, https://www.bmvi.de/SharedDocs/DE/Publikationen/DG/broschuere-strategie-automatisiertes-vernetztes-fahren.html (last visited March 2, 2021).

[274.] Id.

[275.] German Road Traffic Regulations [online] available at https://www.bmvi.de/blaetterkatalog/catalogs/327056/pdf/complete.pdf (last visited March 21, 2021).

is that the AVs have the right technical devices and controls to perform their tasks under the laid down traffic rules and regulations.

The StVG also mandates to have the capability for the automated systems to be taken over by the safety driver in case of any emergent requirement[276]. The StVG also has an ask of designing a manual vehicle control that can be recognised by a safety driver and this system must be capable of alerting the safety driver visually, acoustically, tactilely, or otherwise in such a manner as can be safely perceived by the safety driver. The StVG also mandates that if an alert is raised by the AV, the automated system must ensure the availability of adequate reaction time for the safety driver to take control of the AV.

The most prominent difference between the US regulations and that of Germany is in the definition of AVs. In the USA Level 5 AVs can be driverless. But as per StVG in Germany, there is a need to have a safety driver in all AVs and they must have the capability to take over control on a required basis[277]. The StVG also defines the aspect of being deemed a driver[278]. According to Article §1a (4), anyone who engages in a highly or entirely automated driving function and utilises it to control a vehicle qualifies as a driver, even if he does not manually operate the vehicle at the time the automated function is being used as intended[279]. Article §1b (1) of the StVG includes a feature that allows the driver to divert his attention from the road and the car's controls when it is in fully automatic mode. However, this is subject to the condition that the driver must always remain alert to take over control of the AV if required at any stage[280]. Para (2) of Article §1b, makes it obligatory on the part of the safety driver to take over control of the AV if specifically indicated by the automated system in the AV or if the driver so perceives it to be required basis his visual observation of the various parameters in the vehicle and external surroundings. By this definition, a driver also includes persons who do not exercise physical hand touch control over a vehicle. That's the reason why all road traffic laws in Germany apply to drivers of fully automated vehicles irrespective of their physical presence in the vehicle.

---

276. Id.
277. Id.
278. Id.
279. Id.
280. Id.

The amended StVG law also explains the relationship between the safety driver and the AV. These rules define the rights and responsibilities of the safety driver when engaging the AV in fully automatic mode. Needless to say, these rights and responsibilities of the safety driver are over and above the ones prescribed for the driver of a traditional automobile. Apart from the various provisions regarding the status and responsibilities of a driver in an AV, the StVG also has provisions concerning data and handling of data (§63a). Some of these provisions have been strongly criticised in various write-ups. While using AV quite a lot of data is stored through satellite navigation systems such as the precise moment when the safety driver needs to take over the wheel, the precise location (coordinates) where that moment happens, and the precise moment when the AV malfunctions and the safety driver must take over control.

The provisions concerning data storage and transmission are enumerated in Article §63a (4) of the StVG. The ask in those provisions are that the AV owner is mandated to delete the data which is stored after six months, whenever sought by law enforcement authorities the stored data can be forwarded to those authorities, and lastly in cases where data is forwarded to law enforcement authorities at their request then the owner of the AV must delete the stored data (and transmitted) after three years.

As regards liability concerning AVs even though the German legislature made several amendments to the StVG concerning full AVs, these amendments did not touch upon liability. Hence the liability of a safety driver under German laws would continue to be based on Article §823 of the German Civil Code (BGB)[281] and §18 of the StVG[282].

In addition to the fault-based liability outlined in Article 7 of the StVG[283], the rule of strict liability is also permissible to be applied. This is in sync with the European liability systems where the strict liability rule may be invoked to compensate victims of motor vehicle accidents. There have always been multiple thoughts in Germany on the aspect of the same kind of liability[284]

---

[281] German Civil Code [online] available at https://www.gesetze im internet.de/englisch_bgb/ (last visited March 21, 2021).

[282] German Road Traffic Regulations Supra Note 294, at 75.

[283] Id.

[284] Legal framework of Autonomous driving in Germany [online] available at https://www.researchgate.net/publication/342143289_The_Legal_Framework_of_Autonomous_Driving_in_Germany (last visited March 21, 2021).

being made applicable to AVs. Many have mentioned that German laws must redefine existing liability structures for AVs. However, since as of date there have been no modifications to the existing German laws, the owner of an AV continues to be liable for damages caused by his vehicle as per existing provisions of the StVG. According to Hilgendorf, the existing responsibility structure for partially or completely automated motor vehicles should have been reformed by German legislation[285]. Due to strict responsibility laws under the relevant articles of the StVG, the keeper of a highly or fully automated vehicle is nevertheless accountable for any damage caused by his vehicle because such modification was not adopted. However, while the driver may be released from responsibility if the automated driving feature was on and the harm was brought on by the automated driving system failing, the keeper is not eligible for this release[286].

Despite the above handicap in the existing German laws, once the case is in court the driver and the AV owner can defend the case and get themselves excused from liability if the AV was being used in Automated mode and the damages were caused due to malfunction of the automated driving system. Eric Hilgendorf, a German scholar recommended that for a defect in the product, the manufacturer of the product or the operator be made liable for damages[287]. Additionally, Hilgendorf noted that other laws, such as the German Product Liability Act and the relevant sections of the German Criminal Code, had not been changed. Therefore, in the event of a product defect, the manufacturer or the user of the technological system may be responsible for the harm. König's most recent study highlights the need for laws governing the keeper, driver, and producer's liability and concurs with Hilgendorf's assessment.

The inserted new provisions in StVG are more experimental as can be discerned from Article §1c[288]. According to this Article, after 2019 the responsible ministry (i.e., the Federal Ministry of Transport and Digital Infrastructure) is mandated to revise the amended provisions keeping in mind the grounds of economic considerations. Following the revision, the Ministry is required to inform the Bundestag of the findings.

---

[285]. Id.

[286]. Id.

[287]. Id.

[288]. German Road Traffic Regulations Supra Note 294, at 75.

Because of the foregoing one can safely state that the amended provisions of StVG have helped create a well-defined legal framework for use of self-driving cars on open roads. These rules are unique and exemplary in Europe and hence can serve as a model law for other European states in the process of enabling these states to come up with their regulations on AVs.

2.7.4 Russia

The Russian government, on 26 November 2018 issued regulations that allowed autonomous vehicle testing on Russian roads. This was done vide Regulation no 1415[289]. This was the first step in terms of regulations by the Russian government for AVs. The decision to test the operation of highly autonomous vehicles on public roads in the regions of Moscow and the Republic of Tatarstan was made by the regulation. To facilitate the testing, the Russian government has passed the Rules of Experiment. These rules allow government labs to help in coordinating the testing of self-driving cars or AVs with SAE levels 3 to 5. These labs have the mandate to summarise and submit all test results to the government by 2022 March and provide their recommendations concerning technical requirements and standards for the practical use of AVs. The process followed is that the applicants must submit their application to these labs, obtain permits for testing, and thereafter their test results would be reviewed by these labs, who would then provide safety recommendations post a 45-day review period. As per existing Russian regulations, consumers are not allowed to use AVs with SAE levels 3 to 5 on their public roads[290].

As per Russian regulations, every AV must mandatorily have a data recording system. Apart from this AVs must also be equipped with a mechanism for drivers to activate and deactivate the self-driving system. The equipment used to capture traffic and the driver's activities must also be able to be turned on and off by the driver[291]. The Russian regulations also mandate that every driverless vehicle must have a large black letter "A" mentioned on it to identify

---

[289.] Government.ru. 2021. О проведении эксперимента по эксплуатации на автодорогах высокоавтоматизированных транспортных средств. [online] Available at: <http://government.ru/docs/34831/> [last visited June 28, 2021].

[290.] Id.

[291.] Id.

it as an AV. A minimum of three years of professional driving experience and a spotless driving record are required to be an AV driver. Additionally, the driver must be deemed medically suitable to serve as a driving teacher for the type of vehicle being evaluated[292].

As regards liability, as per Russian regulations, if no one is found guilty of accidents in traffic and other areas on Russian roadways, the AV owner will be liable. All AVs must be insured and protection must be provided for at least 10 million rubles (approximately USD 150,000). As per consumer laws in Russia, consumers can file general liability claims for the warranty of AVs, against the manufacturer. Thus, an AV manufacturer will face similar liability akin to a regular car manufacturer once they start selling AVs to consumers[293]. Russian laws allow a legal entity also to be an owner of an AV and during the testing period, the owner cannot be changed. As per the understanding of Russian lawyers, AV accident disputes can be addressed under general civil law and there is a fair possibility that soon the owners of such driverless vehicles can be made financially liable irrespective of the damage severity. However, there are quite a few in Russia who also have opined that such liability can be placed on the AV drivers who fail to monitor the AV systems or on developers of faulty AV technology, especially in cases of system failure[294].

The initial draft of Russian regulation prescribed that max of 200 driverless cars can be tested at the same time and the capital of the enterprise owning the AV must be not less than US Dollar (USD) 1.5 million[295]. This requirement received a lot of flak as it was a threat to startups and hence was diluted in the final version of the regulations. Yandex is the major driverless technology developer in Russia. Other major international brands like Audi, Toyota, Scania, and Volvo have conveyed their interest in testing AVs in Russia. There is a general industry buzz that by 2025 Russia will have 20000 plus driverless cars on their roads. The main deficiency as can be seen in the Russian rules is that the rules do not spell out the norms to successfully qualify for a test.

---

[292]. Id.

[293]. CEE Legal Matters. 2021. *AVs Regulation in Russia*. [online] Available at: <https://ceelegalmatters.com/cms/13891-autonomous-vehicles-regulation-in-russia> [last visited June 28, 2021].

[294]. AVs Law and Regulations in Russia [online] available at https://cms.law/en/int/expert-guides/cms-expert-guide-to-autonomous-vehicles-avs/russia (last visited March 2, 2021).

[295]. Russian Government begins testing driverless cars [online] available at https://www.loc.gov/item/global-legal-monitor/2019-01-18/russia-government-begins-testing-driverless-cars/ (last visited March 21, 2021).

2.8   Important/landmark case laws on AVs in USA and Russia

2.8.1   *Uber Self-Driving Vehicle Crash in Tempe, Arizona*

Case Summary:

On 18 March 2018, a prototype Uber self-driving Volvo XC90 car struck Ms. Herzberg as she crossed Mill Avenue without using the approved pedestrian crosswalk while pulling a bicycle weighed down with shopping bags. Ms. Herzberg died of injuries in the hospital. The investigations revealed that the speed of the vehicle required an emergency braking system to work. Uber turned off the emergency braking factory settings to lower the possibility for unpredictable vehicle behaviour." National Transportation Safety Board (NTSB) did not apportion any blame to Uber in their preliminary report[296]. As per Uber although the vehicle was officially in autonomous mode, the onboard safety driver was meant to take over in the event of such an emergency, however, this did not occur. NTSB mentioned that there were no problems or diagnostic alerts present at the time of the incident, and the entire self-driving system was functioning normally[297].

■ **Video Footage Analysis:**

The Tempe Police carried out an analysis of dashboard cameras and released two videos displaying video from the two onboard cameras, one of which is forward-looking and the other which is used to capture the actions of the safety driver[298]. NTSB also mentioned that since Ms. Rafaela Vasquez, the safety driver, did not step in to stop the collision, it could not be avoided. Ms. Rafaela only responded by shifting the steering wheel just before the collision and applying the brakes immediately after impact while spending 34% of her time viewing a television show on her smartphone. The car was moving in the far-right lane when the collision happened, according to the front-facing camera. Before the collision, the safety driver was seen gazing

---

[296.] Andrew J. Hawkins, Uber self-driving car saw pedestrian but didn't brake before fatal crash, feds say The Verge (2018), https://www.theverge.com/2018/5/24/17388696/uber-self-driving-crash-ntsb-report (last visited December 30, 2020).

[297.] Id.

[298.] Id.

down by the driver-facing camera.[299] When questioned, the driver defended herself by saying that at the moment of the crash, she was monitoring the centre stack.

- ■ **Software Issues:**

The vehicle software did have issues. The instrument recording or telemetry recording revealed that the victim was detected only six seconds before the crash. The victim was initially identified as an unidentified object before being identified as a car and subsequently as a bicycle. The autonomy logic had several paths anticipated for each of these three identities.

- ■ **Sensor issues:**

Advanced sensors on the car included radar and light detection and ranging (LiDAR) sensors. These sensors are not normally affected by darkness and hence it was safely concluded that "there were no sensor issues and if at all there was any issue it was the safety driver not taking control when required".

- ■ **Actions by Uber post the Accident:**

Uber stopped testing autonomous vehicles in Arizona after the fatal event and then resumed it nine months later. At the end of March 2018, when its permit to test driverless vehicles in California expired, Uber made the decision not to renew it. In February 2020, the California Department of Motor Vehicles re-issued the permit afresh[300] to Uber Advanced Technologies for testing of AVs and Uber thereafter started testing in San Francisco. Uber in collaboration with Volvo has now come up with the third generation of Autonomous cars which has a lot of redundancy, specifically around steering, braking, and battery power[301]. If any of the major systems should fail for whatever reason, the backup system is intended to immediately act to halt

---

[299] Preliminary Report Highway HWY18MH010, https://www.ntsb.gov/investigations/AccidentReports/Reports/HWY18MH010-prelim.pdf (last visited December 30, 2020).

[300] Uber gets permit to restart testing its self-driving cars in California, REUTERS (2020), https://www.reuters.com/article/us-uber-self-driving/uber-gets-permit-to-restart-testing-its-self-driving-cars-in-california-idUSKBN1ZZ2QG (last visited December 30, 2020).

[301] Sean O'Kane, UBER DEBUTS A NEW SELF-DRIVING CAR WITH MORE FAIL-SAFES THE VERGE (2019), https://www.theverge.com/2019/6/12/18662626/uber-volvo-self-driving-car-safety-autonomous-factory-level (last visited December 30, 2020).

the car, according to Uber. NTSB had in one of their reports blasted Uber for poor safety culture. Uber thereafter also stated that "going forward Uber will not disable the automatic braking system"[302].

■ **Current Status of the Case:**

Post investigation NTSB divided the blame between Uber, the safety driver, the victim (apart from crossing the road at a non-designated place also had tested positive for methamphetamine and marijuana), and the State of Arizona. The NTSB report mentioned the failure of the federal government to properly regulate the AVs industry. Uber was vindicated of any criminal wrongdoing by local authorities. As regards the lawsuit filed by Herzberg's family, Uber managed an out-of-court settlement for an undisclosed sum[303]. After being charged on August 27, 2020, Ms. Vasquez made her initial court appearance on September 15, 2020. The trial was expected to commence in February 2021, but it got pushed to 11 May 2021 and thereafter to 10 August 2021[304]. As of July 2023, the latest update on the Elaine Herzberg case involving an Uber self-driving car is that the backup driver, Rafaela Vasquez, pleaded guilty to endangerment and was sentenced to three years of supervised probation and must pay restitution to all victims, including insurance companies. Vasquez was originally charged with negligent homicide in 2020 but pleaded guilty to the reduced charge of endangerment in July 2023.

Issues having no clear answer in the investigation so far

Even though the names of Arizona State and Uber were listed in the initial investigations, no final charge-sheet has been filed against them. The only case in the court was against the safety driver. Uber managed to close the case filed by the relatives of the deceased by compensating them for an undisclosed sum.

---

302. Id.

303. Andrew J. Hawkins, Uber has resumed testing its self-driving cars in San Francisco The Verge (2020), https://www.theverge.com/2020/3/10/21172213/uber-self-driving-car-resume-testing-san-francisco-crash (last visited December 30, 2020).

304. Trial Delayed for Backup Driver in Fatal Crash of Uber Autonomous Vehicle [online] available at https://www.phoenixnewtimes.com/news/uber-crash-arizona-vasquez-herzberg-trial-negligent-homicide-charge-11553424 (last visited February 23, 2022)

Under normal circumstances, one would have expected Uber and State of Arizona to have been made responsible under the Absolute Liability doctrine (akin to the MC Mehta & Another versus Union of India in the Bhopal gas case reported in 1987 AIR 1086). Uber for negligence (jeopardizing public safety) by choosing to disable the lifesaving feature in such a technology which may endanger human lives and the State of Arizona for not having proper clarity in the law/policy related to testing of AVs. Surprisingly, this did not happen. Also, there seems to be no Product Liability case filed against Uber/ Volvo/anyone the software/sensor vendors of the AI systems used in the vehicle involved in the crash. That's the second surprise exclusion.

2.8.2   *Tesla Autonomous Car Crash cases*

As of date, there have been six fatal car crashes involving Tesla cars with autopilot of which three have been in the USA.

2.8.2.1   *Williston Florida Tesla Crash*

The first U.S. Autopilot fatal accident involving Joshua Brown, 40, of Canton, Ohio, occurred on 07 May 2016 at Williston in Florida. According to NHTSA, the tractor-trailer turned left in front of the Tesla car, causing the collision. Evidence shows that the driver failed to use the brakes when the collision occurred at an intersection on an uncontrolled access route. After the incident, the car kept going until it passed beneath the truck's trailer and then struck a telephone pole a quarter mile ahead. During the post-crash inspection of the wreck, the Florida Highway Patrol discovered a portable Digital Video Disc (DVD) player and a laptop computer. In addition, an adaptable laptop mount for vehicles fastened to the frame of the front passenger seat was also found[305]. There are reports which also state that the sensors did not identify the white tractor against a bright sky. Tesla claims that neither the autopilot nor the driver noticed the white side of the tractor-trailer against a brightly lit sky, hence no brake was deployed. The driver of the truck claimed to have heard a Harry Potter movie playing inside the wrecked vehicle.

[305.] Tesla Autopilot, WIKIPEDIA (2021), https://en.wikipedia.org/wiki/Tesla_Autopilot#Handan,_China_ (January_20,_2016) (last visited January 2, 2021).

The NHTSA in their report released in January 2017 mentioned that to see the vehicle, the driver had seven seconds. The NHTSA also confirmed that the Autopilot software was faultless. The report[306] concluded that since the vehicle was in testing mode, despite the autonomous technology available, the final responsibility for retaining control of the vehicle rested with the driver and he failed to do so. According to the NHTSA, the driver may have been preoccupied at the time of the crash and the laptop was presumably mounted. The NTSB released its report in September 2017 and concluded that the Williston, Florida crash was likely caused by the truck driver's failure to give way to the car, combined with the car driver's inattention from over-dependence on vehicle automation, which led to the car driver's lack of reaction to the truck's presence. According to the NHTSA report, before the collision, the driver had been using Autopilot for about 37 minutes. Thirteen hands not detected signals had been sent by the automated system, and the driver had replied to each one on average sixteen seconds later.

Legal Aspects:

All Tesla vehicles carry a warning which indicates that autopilot is disabled by default. The safety drivers have to acknowledge that they understand that as a novel technology, the autopilot must first pass a public beta test before being made available for commercial usage. Despite this, the legal experts felt that in this case Brown's family still had a legal case to indicate that the driver could have been led to rely heavily on the system's capability. There are even other attorneys who feel that the family of Brown has a product liability case against Tesla basis the fact that Brown was not appropriately warned about the likely defects which could crop up in the autopilot system. Tesla also responded in their blog mentioning that the car has an automated system to remind the safety driver to always keep their hands on the steering. In addition, Tesla mentioned that the car's AI system also carries out regular checks by giving visual cues and reducing speed until hands are seen, to make sure the driver retains his or her hands on the wheel. The alerts did happen in this case as the NHTSA reports state. All these indicated that Tesla would have pointed to the driver's actions in any legal case by mentioning that it was the driver who accepted /assumed the risk making him fully or partially culpable. Tesla

---

[306]. Id.

also would have showcased the warnings provided by them on prior use/beta version requiring supervision and stated that the acts of the driver were the primary cause of the crash. Maybe because of all these considerations the family of Mr. Brown did not sue Tesla. It is also not known if Tesla privately compensated them[307].

2.8.2.2 Mountain View, California Tesla Crash

The second US Tesla autopilot fatal crash occurred on 23 March 2018 in Mountain View, California. The crash occurred in the morning hours at the carpool lane exit for highway 85 when the car hit a concrete barrier and as a result, two more following vehicles crashed behind it[308]. Thereafter the car caught fire. The NHTSA investigations show that "the driver Walter Huang had five seconds reaction time and 150 metres distance available for reacting, but he did not".

There have been rumours that the autosteer feature accidentally marked the far-left lane's lane with the left-side white line rather than the right-side white line. In February 2020, the NTSB concluded that the driver's excessive dependence on the autopilot system, its limitations, and the likelihood that he was playing a video game on his phone all played a role in the collision. The faded left line also probably contributed to the crash. Since NTSB is only a recommendatory body it forwarded its recommendations to two regulatory agencies. There is no publicly available news on any compensation paid by Tesla to Mr. Huang's family nor is there any news on any product liability case against Tesla.

2.8.2.3 Delray Beach, Florida Tesla Crash

On 01 March 2019, a Tesla Model 3 hit a semi-trailer truck that was turning left. The impact was so huge that the greenhouse of the Model 3 was sheered, and the Tesla driver Jeremy Banner died[309]. Investigations revealed that the

---

[307] Tesla fatal autopilot crash: family may have grounds to sue, legal experts say, THE GUARDIAN (2016), https://www.theguardian.com/technology/2016/jul/06/tesla-autopilot-crash-joshua-brown-family-potential-lawsuit (last visited January 2, 2021).

[308] Tesla says Autopilot was active during fatal crash in Mountain View [online] available at https://arstechnica.com/cars/2018/03/tesla-says-autopilot-was-active-during-fatal-crash-in-mountain-view/ (last visited January 2, 2021).

[309] Tesla's Autopilot was engaged when Model 3 crashed into truck [online] available at https://www.theverge.com/2019/5/16/18627766/tesla-autopilot-fatal-crash-delray-florida-ntsb-model-3 (last visited January 2, 2021).

Tesla driver had engaged the autopilot just ten seconds before the crash. Also, the telemetry results did not reveal that before the collision, the driver had his hands on the wheel for eight seconds. Also, the 2019 NTSB report revealed that "no evasive manoeuvres were executed by either the Tesla or the Truck driver"[310]. The NTSB also emphasised that it had encouraged Tesla to only utilise autopilot on limited access highways after the Florida Tesla disaster, but that Tesla disregarded this advice.

Legal Aspects:
The family of Banner filed lawsuits against Tesla, First Fleet Trucking, and semi-truck driver Richard Keith Wood. According to the Palm Beach County cap for filing a civil lawsuit, the family requested damages totaling more than $15,000. Tesla representatives said, in the ten seconds before the disaster, Banner only utilised the autopilot once. As per Tesla autopilot when correctly employed by a careful motorist who is constantly ready to take control is safer than those operating with assistance. This has been Tesla's stand in most cases.

2.8.2.4   *Non-Fatal Tesla Crashes*

Other than the above fatal crashes two more non-fatal crashes of Tesla have occurred in the past in the USA and one in Russia. The first one was on 22 January 2018 in Culver City, California wherein a Tesla Model S hit a stationary fire truck after the vehicle ahead of it changed lanes and the Tesla's sensor/autopilot failed to detect the stationary truck in time[311]. NTSB also pointed out that "post the Williston crash it had advised Tesla to have in place appropriate sensors to detect driver's activity and alert the driver if any kind of engagement was lacking. That was not taken care of by Tesla"[312].

The autopilot collision warning came up only half a second before the crash but did not engage the automatic braking system. There are many reports which indicate that "the autopilot is not as good in detecting stationary objects as it can detect moving objects". The NTSB decided that the driver was at fault in its report, which was submitted in August 2019, and based its

---

[310.] Id.

[311.] Tesla car was on Autopilot when it hit a Culver City firetruck, NTSB finds [online] available at https://www.latimes.com/business/story/2019-09-03/tesla-was-on-autopilot-when-it-hit-culver-city-fire-truck-ntsb-finds (last visited January 2, 2021).

[312.] Id.

conclusion on the driver's inattention and excessive reliance on the vehicle's advanced driver assistance system[313].

The second non-fatal crash occurred in South Jordan, Utah on 11 May 2018 again with a Tesla Model S which hit a stationary truck. In this case, the driver of the Tesla was injured in her foot, and she admitted to looking at her phone just before the crash[314]. According to Heather Lommatzsch of Lehi, who filed the lawsuit, she understood from conversations with Tesla salespeople that the Tesla Model S's safety features would ensure the car would stop on its own if an obstacle were to cross its path, but in her accident, these safety features did not function as promised. Additionally, Lommatzsch asserts that a Tesla dealer advised her that she could operate the vehicle in autopilot mode with only occasional steering wheel inputs. NHTSA reports indicate that the driver did not touch[315] the steering wheels at least for eighty seconds before the collision, and the driver was held responsible for the mishap despite the driver's claims that the vehicle failed to issue any audio or visual warnings before the collision. That did not happen due to the differential handling by autopilot of stationery objects vis-à-vis moving objects.

On the evening of August 10, 2019, a Tesla Model 3 travelling in the left lane of the Moscow Ring Road in Moscow, Russia collided with a tow truck that was stopped in the lane with a corner sticking out before catching fire. The driver asserted that the car was moving at the legal limit of 100 kilometres per hour (62 miles per hour) while using Autopilot and that he was not paying attention even though his hands were on the wheel. Before the car caught fire, every passenger was able to get out, and they were all taken to the hospital. A driver's leg was shattered, and his children had bruises[316]. A surveillance camera caught the tow truck being pushed against the centre-dividing wall by the collision's impact. Several films of the fire and explosions that followed the collision were also recorded by bystanders. These videos also show that

---

313. Id.

314. Driver in South Jordan auto-pilot crash sues Tesla [online] available at https://www.deseret. com/2018/9/5/20652799/driver-in-south-jordan-auto-pilot-crash-sues-tesla (last visited January 2, 2021).

315. Id.

316. Tesla Explodes After Crashing into Tow Truck While on Autopilot [online] available at https://www.newsweek.com/video-tesla-explodes-tow-truck-1453745 (last visited March 21, 2021).

the tow truck that the Tesla crashed into had been moved, indicating that the explosions of the Model 3 occurred afterwards.

2.9 Conclusion

According to the National Safety Council statistics, around 40,000 people die due to road accidents in the USA and apart from these deaths millions of injuries have also occurred due to road accidents. The medical attention costs have been close to the tune of around 400 million plus dollars per year. The National Highway Traffic Safety Administration (NHTSA) carried out a study and it proved that human error has been the reason for more than 90% of the deaths caused due to road accidents. The NHTSA then concluded that "an autonomous vehicle which can be programmed to drive on its own without any kind of error was probably the best solution to put an end to over 40,000 humans being killed by humans each year in the USA". Insurance companies are also supporting this, and they have started offering discounts to drivers of vehicles with safety devices.

Up to November 2020, thirty-eight states and the District of Columbia (DC) have passed laws/executive orders for testing and plying of AVs (AVs)[317]. Five states have only authorized studies to define the framework for AVs and to look into various funding aspects for the research, development, and infrastructure related to plying of AVs. Twelve states have permitted testing. Sixteen states and DC have permitted the full deployment of AVs. Among the aforementioned, 18 states currently permit testing and deployment without a safety driver in the car, while 4 states have regulations on truck platooning. Other states either have them in the pipeline or are yet to decide on them. AVs are rated according to their automation levels. The basic standard is six levels of automation. All manufacturers must carry out a safety assessment as per the Department of Transport, USA norms. Irrespective of the level of automation there is no excuse from obtaining vehicle registration, insurance, and driving license. Different levels of automation are categorised using a zero–five rating scale developed by the Society of Automotive Engineers in the USA (SAE).

In 2017 the US House of Representatives enacted the Safely Ensuring Lives Future Deployment and Research in Vehicle Evolution Act (SELF DRIVE

[317.] AVs, GHSA, https://www.ghsa.org/taxonomy/term/521 (last visited January 10, 2021).

Act) vide H.R. 3388[318]. This bill was primarily aimed at adopting federal safety standards in the arena of AVs. The bill sought of attempted to clarify the roles of federal and state governments concerning this emerging technology. A few weeks after the passing of this Act, a similar bill called American Vision for Safer Transportation through Advancement of Revolutionary Technologies Act (AV START Act) under reference S. 1885 was also presented, but never passed.

The draft language of the AV Start Act is very similar to the SELF-DRIVE Act and incorporates the creation of a Highly Automated Vehicle Advisory Council within the Department of Transportation to assess problems with autonomous vehicle concerns and deal with federal regulation surrounding AV testing. Under the AV Start Act, the onus was vested in each AV manufacturer to provide a written plan for identifying and reducing cybersecurity risks. Both bills finally expired when the new Congress came into power in 2018. Two Republican Senators reintroduced the Self-Drive Act again on September 23[319]. The Senators have mentioned while introducing the bill that "this is aimed towards creating a national framework for the development and testing of AVs or AVs".

To facilitate the implementation of universal approval standards for motor vehicle equipment and parts in Europe, the UNECE Agreement was developed in 1958. Thereafter the 1988 Global Technical Regulations (GTR) Agreement[320] was introduced for creating a global process for this type of approval, applicable to all countries in the world. EU also adopted a Directive that specified that once a vehicle or a component of a vehicle is certified in one Member State, it cannot be banned from other States' Markets. Should there be a need to exclude them, it can only be done with valid reasoning. To appropriately allocate risk associated with the use of motor vehicles, the EU has also developed many tools. The Product Liability Directive and the Motor Insurance Directive are the two primary pieces of legislation that govern liability.

---

[318] Congress Taking Another Look at Regulating Automated Driving Systems, JD SUPRA, https://www.jdsupra.com/legalnews/congress-taking-another-look-at-51054/ (last visited December 17, 2020).

[319] Republican House Members Reintroduce SELF DRIVE Act, AASHTO Journal (2020), https://aashtojournal.org/2020/09/25/republican-house-members-reintroduce-self-drive-act/ (last visited December 17, 2020).

[320] UN Vehicles Regulation Agreement [online] available at https://unece.org/DAM/trans/doc/2018/UNDA/III_UN_vehicle_regulations_agreements.pdf (last visited April 12, 2021).

Germany has since 2017 permitted up to level 3 automation and in some cases up to level 4 automation[321]. This was done to provide more clarity to the automobile industry and security for investments. The said German legislation lays down the conditions when the safety driver has to intervene and if there is any lapse then the driver is held responsible in case of an accident.

The analysis of the existing legal regime in the USA and developed nations of the EU concerning AV indicates that all these countries have a robust and well-planned scheme to permit plying of AVs in their countries. These countries have very clearly defined who is an Autonomous Vehicle driver, what is an AV, how much insurance is mandatory before testing an AV, and how much when plying an AV. Since truck platooning is more popular in the USA, most states in the USA have specific laws for truck platooning and they have exempted the close distance rule for succeeding vehicles in the case of truck platooning this sounds logical as all vehicles in a platoon are controlled by the leading vehicle. Also, since the number of AV accidents has been high in the USA, number-wise, they have been learning from past accidents, and accordingly, most states are continuously tweaking their laws, especially concerning the safety of vehicles and pedestrians.

* * *

[321.] Germany will be the world leader in autonomous driving [online] available at https://bmdv.bund.de/SharedDocs/EN/Articles/DG/act-on-autonomous-driving.html, (last visited April 12, 2021.

CHAPTER 3

LEGAL REGIME IN ASIA AND THE REST OF EUROPE FOR AUTONOMOUS VEHICLES

3.1 Background of Autonomous Vehicles in Asia and the Rest of Europe

Autonomous vehicles or AVs are the most happening progress in the automobile sector. As a result, there is a great amount of time and investment being spent by most nations on Research and Development (R&D) in this fast-growing sector. Reports are stating that Asia Pacific Japan China (APJC) region will probably have 24 million AVs by 2024. Apart from AVs another big thing on which nations are focusing is climate change and towards this Electric Vehicles or EVs are also an important aspect of R&D[322]. According to the survey, the region's widespread usage of ADAS technology is promoting the expansion of the AV sector. To improve their self-driving solutions, several automakers are forming strategic alliances with tech firms. In addition, partnerships are being established amongst automakers themselves to develop such technology[323]. There is a great scope to integrate AV with EVs and many nations across the world are working in that space too. In the APJC region, many countries are

---

[322.] www.ETAuto.com, Autonomous cars to grow to 24 million units in Asia-Pacific by 2024 - ET AUTO ETAUTO.COM (2019), https://auto.economictimes.indiatimes.com/news/auto-technology/autonomous-cars-to-growth-to-24-million-units-in-asia-pacific-by-2024/70701274 (last visited April 25, 2021).
[323.] Id.

focusing on building appropriate infrastructure for EVs, which could propel the growth of AVs[324] too. Some of the major players are Volvo, Ford, Honda, General Motors, Volkswagen, and Toyota. Apart from these companies, Uber is also involved as a shared mobility solution provider Honda and GM's subsidiary Cruise have a deal to create new autonomous vehicles[325]. $2 billion in investment is expected over 12 years. As a beginning, the Joint Venture (JV) has made an immediate investment of $750 million. Singapore and Malaysia are pioneer nations in this regard.

Many automobile manufacturers in APJC and ANZ are collaborating with a broad spectrum of tech companies to promote advancement in their AV solutions. Apart from this, these auto companies are also mutually collaborating amongst themselves to control costs and bring about better efficiency.

As in the USA, manufacturers and AI technology innovators in Europe are working towards bringing out AVs for consumer use. However, when this will become a mass reality is a question for which there is no concrete answer. The dedication in all countries in Europe can be seen as each is working towards becoming an AV leader. Even governments across many countries in Europe are supporting these efforts through research funding and appropriate legislation to permit trials. Most nations in Europe have also realized that for smooth implementation of AVs they will have to align their AV programs with other countries and try to make them as uniform as possible. Some countries in Europe are at an advanced stage of testing AVs while many are still in the preliminary stages only.

Around 1942 Isaac Asimov in his three robotic laws propounded the relationship between humans and technology[326]. When Asimov said that the creators of autonomous technology must explicitly assign responsibility at the outset itself, there was a great deal of insight in his statement. Asimov's proposed regulations were created to safeguard people against encounters with robots. They are: (a) A robot may neither harm a human being or, by doing nothing, let a human suffer harm; (b) A robot must obey commands from

---

324. Id.

325. Id.

326. After 75 years, Isaac Asimov's Three Laws of Robotics need updating [online] available at https://theconversation.com/after-75-years-isaac-asimovs-three-laws-of-robotics-need-updating-74501 (last visited March 02, 2021).

humans unless doing so would violate the First Law; and (c) As long as it doesn't break the First or Second Laws, a robot must defend its existence.

3.2 Approach by various countries to support AVs

Autonomous Vehicles which were a dream till a few years back are now slowly turning into reality. While working towards regulating AVs the lawmakers have to work efficiently towards the correct mapping of all applicable legal issues. Liability would be the top question when one talks of AVs. This would be closely followed by questions relating to Intellectual Property (IP). The software used in AVs would trigger copyright laws. Similarly, the Blackbox used to record data in the AVs would raise questions connected to privacy and data protection.

A 100 square kilometre enclosed area has been made operational by Shanghai to test AVs[327]. The base is working to establish itself as China's primary facility for the development of standards and specifications, testing, and certification for cutting-edge ICV technology. It also intends to develop into a hub for talent development, industry gathering, and ICV innovation. This is going to be called the National Intelligent Connected Vehicle Pilot zone[328]. This pilot zone has a lot of sensors and communication modules to facilitate the testing of AVs. Some of the brands which participated in the first round of pilot testing are Ford, Volvo, Cadillac, and Volkswagen[329]. The Pilot Zone also hopes to develop into a hub for talent development, industry gathering, and ICV innovation. The local and international car makers of these brands and their technology suppliers sat in these test vehicles behind the steering to monitor the AVs. The Chinese Academy of Science, Tsinghua University, Tongji University, and Shanghai Advanced Research Institute sent experts on these test drives as well[330].

China's 2025 vision includes Intelligent Connected Cars as a key strategic priority[331]. Jiading is one of the top districts in China which is well known

---

[327] Autonomous connected cars on their way, archive.shine.cn (2016), http://english.jiading.gov.cn/2020-04/09/content_37535207.htm (last visited April 25, 2021).

[328] Id.

[329] Id.

[330] Id.

[331] China wants self-driving tech in half of new cars by 2025 [online] available at https://asia.nikkei.com/Business/Automobiles/China-wants-self-driving-tech-in-half-of-new-cars-by-2025 (last visited March

for the promotion of innovation in the automobile industry. During 2016, this district nurtured a total of thirty-one automobile sales units and around twenty-eight thousand cars were sold[332]. As per Chinese experts, AV driving in China is more complicated than in the USA and hence if AVs achieve success in China, they can easily ply in the USA[333]. China's intelligent linked vehicles will have a fully developed industrial system, enhanced industrial ecology, and intelligent level of connected vehicles by 2035.

Singapore-based NuTonomy, an initiative of MIT has started a self-driving taxi service covering two and half square miles. There are more than six AV taxis available in this area. These operate at predetermined pickup and drop-off points. The cabs are mainly from Renault and Mitsubishi[334]. In addition to the above Delphi also operates six self-driving Audi SQ5s in an area of five square miles. The Delphi taxis require a safety driver, but the aspiration is to remove the safety driver completely at a later date.

As discussed in earlier chapters GM made the first driving car as an exhibit in 1939 and followed it up in 1958 by making it a reality. This was improved by Japan in 1977.[335] The Japanese created this concept in 1977 by utilising a camera system that transmitted data to a computer so it could decipher photographs of the road. However, this vehicle could only go at a speed of under 20 mph[336]. Toyota from Japan has also been carrying out AV trials on the Shuto Expressway in Tokyo. They have been using a modified Lexus[337] GS for these trials. German engineers made advancements ten years later with the VaMoRs, a camera-equipped vehicle that could safely travel 56 mph. The capability of self-driving cars to notice and respond to their surroundings has evolved along with technology. The testing by Toyota has mainly been on the highways post on ramp till the car is off-ramp. These AVs have been able to

---

25, 2021)

[332] Jiading's auto industry highlights innovation [online] available at http://www.chinadaily.com.cn/m/shanghai/jiading/2017-03/28/content_28703451.htm (last visited March 25, 2021)

[333] Roadmap lays out path for connected vehicles [online] available at https://www.chinadaily.com.cn/a/202011/16/WS5fb1d9c1a31024ad0ba94473.html (last visited March 25, 2021).

[334] MIT spinout NuTonomy just beat Uber to launch the world's first self-driving taxi [online available at https://techcrunch.com/2016/08/24/mit-spinout-nutonomy-just-beat-uber-to-launch-the-worlds-first-self-driving-taxi/ (last visited April 25, 2021).

[335] History of the Autonomous Car, TitleMax (2020), https://www.titlemax.com/resources/history-of-the-autonomous-car/#:~:text=History%20of%20Autonomous%20Cars,made%20this%20concept%20a%20reality. (Last visited April 25, 2021).

[336] Id.

[337] Id.

demonstrate that they can accelerate, brake, and steer by themselves on the highway. Toyota has been mentioning that they intend to reduce accidents through the use of AVs[338].

There is also a conversation that the AV trials are happening to help Japan to use these driverless vehicles for the Olympics which got moved from 2020 to 2021. Toyota e-Palette was used to ferry athletes at the 2021 Japan Olympics and Tokyo Paralympic games. During the Tokyo Paralympic games after one such AV collided with a visually impaired athlete, the running of self-driving AVs was stopped[339]. After a self-driving bus ran over a Paralympic judoka in the athletes' area, Toyota apologised for the overconfidence of the vehicle and announced that the service will be temporarily suspended. The Japanese project aims to build a multi-layered one-stop map containing information that is both static and dynamic and is necessary for the safe travel of AVs.

As regards Australia, Bosch in the year 2016 did Joint Venture (JV) with the Government of Victoria to showcase an AV built using components from Bosch and a Tesla body. This AV took nine months to build, and more than forty-five humans were involved. This AV does require a safety driver even though it can navigate on the road without driver support. Australian roads are famous for Kangaroo accidents. Every year more than twenty thousand Kangaroos are killed in road accidents in Australia and the insurance claims have always been beyond $Seventy-five million. Hence that's also a big concern when having AVs ply on Australian roads. Volvo has been studying this aspect specifically to help them build the right kind of software for AVs, to help detect kangaroos on the road and avoid colliding with them. This will help both the insurers and drivers and also protect Kangaroos[340].

The creation of a uniform EU regulation for AVs will take time and one can expect that to be a gradual and slow process. Many countries in the

---

[338]. Accenture Insurance, Around the world in self-driving cars: Asia Accenture Insurance Blog (2017), https://insuranceblog.accenture.com/around-the-world-in-self-driving-cars-asia (last visited April 27, 2021).
[339]. Toyota pauses Paralympics self-driving buses after one hit visually impaired athlete [online] available at https://www.theguardian.com/technology/2021/aug/28/toyota-pauses-paralympics-self-driving-buses-after-one-hits-visually-impaired-athlete (last visited April 27, 2021).
[340]. Effective Vehicle-Based Kangaroo Detection for Collision Warning Systems Using Region-Based convolutional networks [online] available at https://www.ncbi.nlm.nih.gov/pmc/articles/PMC6022154/ (last visited, March 26, 2021).

EU are slowly and steadily working on regulations/laws for AVs. The AV manufacturers are putting countries under pressure to come up with laws for AV testing and operation on roads. Amendment of existing motor vehicle rules seems to be the common track adopted by most nations.

3.3 Advanced Driver Assistance Systems (ADAS)

Advanced driver-assistance systems (ADAS) as the name suggests use advanced technologies in vehicles to help the driver. These could include many safety features. ADAS could also use sensors in a vehicle like radars and cameras and then analyse the collected information and provide it to the driver if there is a safety driver in the AV else take automated actions as per the inbuilt software program[341]. The intent of ADAS is to increase consumer confidence and boost autonomous driving technology. ADAS helps to notify the driver of any object that could collide like a cycle, a human, another vehicle, etc. ADAS can also take care of blind spots which a human eye cannot. Similarly, if the vehicle is drifting from its lane, again ADAS will come to the AV's rescue.

ADAS also functions as an active safety system. It can efficiently and actively control an AV's braking or steering. That's one of the reasons why the word assistance is included in ADAS. All the above features help ADAS save human lives. Studies have shown that collision warning systems have reduced crash rates by 27% and if automatic braking was included this number is doubled. Similarly, rear-view cameras help reduce crashes while reversing by 17%, and including automatic braking while reversing increases this number to 78%[342].

Another distinguishing feature of ADAS is the adaptive cruise control which aids a car in maintaining a safe distance from the car in front of it[343]. ADAS also can have sophisticated features like steering and propulsion control under certain conditions like highway driving, without the need for hands-on control by a safety driver. Under the SAE system, these fall under the level 2 category and are generally available in the market today.

---

[341] 2020 Mobility Insider October 02, What Is ADAS? Aptiv, https://www.aptiv.com/en/insights/article/what-is-adas (last visited April 28, 2021).

[342] Id.

[343] Id.

In ADAS the driver technically is engaged in driving. The driver's hands need not necessarily be on the steering and maybe they could rest their feet away from the gas and brake pedals. Despite all this, the driver is still responsible to ensure that the AV operates safely. Most OEMs. have installed driver monitoring systems to ensure that the driver is focused on the vehicle at all times even if the vehicle is in automated mode. Many companies are using Satellite Architecture to utilize sensor-detected intelligence to increase performance, reuse, and update ability. This also helps lower costs. One thing is for sure, ADAS will help reduce vehicle-related injuries and fatalities to a great extent.

3.4 Existing laws, regulations, and policies on Autonomous Vehicles in APJC

3.4.1 China

As early as 2015, Baidu which is the company that runs the main search engine in China carried out an AV test. However, AV tests were first officially allowed by the Chinese Government only in 2018 as by then China had a large number of local AV companies[344] that were working hard to compete with global OEMs. The initial AV tests in China were conducted by two Chinese start-ups Jingchi and Pony.ai. These tests were carried out in Guangzhou in 2018[345]. It is said that since Chinese regulators prefer to let businesses experiment with their innovation before intervening, the Chinese environment for testing innovations is easier as compared to other markets in the world. Roadstar is teaming with Renovo.auto, a Silicon Valley company that creates the operating system for self-driving cars, to connect its software to lidar, cameras, and other sensors[346]. The operating system communicates the information the hardware gathers to Roadstar's computer brain, which uses it to control the vehicle. Additionally, Renovo can link Roadstar's vehicles to mapping software and even a ride-hailing service like Uber or Lyft[347].

[344.] To Find China's Best Driverless Technology, Look in Silicon Valley [online] available at https://www.bloombergquint.com/business/to-find-china-s-best-driverless-technology-look-in-silicon-valley (last visited April 25, 2021)."

[345.] Id.

[346.] Id.

[347.] Id.

Although China has exhibited relative improvement in technology innovation, it scores low when it comes to the availability of the latest technology and consumer adoption specifically concerning AVs. When it comes to patents China is behind US and Japan. The expectation is that AVs will become popular in China slowly on certain routes, streets, and highways. Many companies in China focus on AV and AV-related components of which the famous ones are RoboSense and SureStar[348]. These two companies work on lidar. There is another company called Horizon Robotics, which is mainly in the area of AI microchips. With 21 percent of the population having used an online ride-hailing service, China scores highest on a new metric that focuses on this industry, showing relative improvement in the infrastructure, innovation, and technology pillars. However, it performs worse on metrics like the accessibility of cutting-edge technologies and consumer adoption of technology.

As regards motor vehicle laws there are no officially released standards in China (PRC) akin to the standards of the Society of Automotive Engineers (**SAE Levels**). The Chinese government on 09 March 2020 released the draft National Standards for AVs[349]. This Standard classifies AV driving on a pattern just like SAE levels. This Standard was to be tentatively approved on 01 January 2021 but the same has been delayed. As of date consumers cannot use ICVs on public roads in China.

On 03 April 2020, a circular was issued by three Ministries of the PRC government for Road Testing of Intelligent Connected Vehicles or ICVs. As of date, this ICV Road Test Norms (ICV Norms) is the only existing law in PRC for AVs and this is effective from 01 May 2018[350]. The ICV Norms segregate AV driving into three categories: conditional, highly automated, and fully automated. ICV road testing statutory requirements apply to all three categories.

As of date, the ICV testing is permitted only on select roads in PRC. Some of the mandatory conditions[351] are that the person who applies for

---

[348]. 2019 Autonomous Vehicles Readiness Index (2019), https://assets.kpmg/content/dam/kpmg/xx/pdf/2019/02/2019-autonomous-vehicles-readiness-index.pdf (last visited April 29, 2021).

[349]. Global guide to Autonomous Vehicles 2021 [online] available at https://www.businessgoing.digital/global-guide-to-autonomous-vehicles-2021/ (last visited, March 25, 2021)

[350]. Autonomous Vehicles Law and Regulations in China [online] available at https://cms.law/en/int/expert-guides/cms-expert-guide-to-autonomous-vehicles-avs/china (last visited March 2, 2021).

[351]. Id.

the testing has to have a registered legal status in PRC (Entity) concerning ICV, for manufacturing, testing, and R&D. The Entity must have sufficient financial capability to pay up any kind of compensation due to personal or property damages caused by ICV testing. The Entity must have the capacity to conduct, tests, do distance monitoring and carry out recording and analysis of these tests, and if required even reproduce these tests. The driver needs to possess certain prescribed personal qualifications. The ICV being tested must possess requisite technical requirements and must mandatorily have a compulsory traffic accident liability insurance coverage of over Renminbi (RMB) 5 million per tested vehicle.

As of date, there are no ICV-specific insurance laws in PRC and hence in case of any accident, the driver will be liable as per existing laws. In case of any accident during the testing of ICV, compensation for damages shall be determined as per existing laws. Some of the broad guidelines/rules[352] as per existing PRC laws state that when two motor vehicles are involved in an accident, the person at fault is responsible, and where the fault is shared by both parties, liability is determined on a proportionate basis[353]. In case of accidents between a vehicle and a non-vehicle say a pedestrian, the liability is always on the driver unless the driver can prove that he/she was not at fault. Whenever there is proof that a pedestrian or other non-vehicle was also at fault, then the liability is to be proportionately allocated based on the degree of fault. If it is proved that the driver was not at fault, then his compensation liability shall not exceed 10%. Lastly, if evidence indicates that the accident occurred purely owing to the error of the non-vehicle/pedestrian then the driver shall not be liable for any compensation.

Akin to other countries PRC also has a third-party (TP) liability insurance system. This TP insurance has to be as per law, procured by the vehicle owner or user. Noncompliance in this regard can result in the seizure of the vehicle and a fine equivalent to twice the premium for a minimum liability limit[354]. Concerning manufacturer's liability for ICVs, present PRC law has no specific provision, and hence warranty claims will have to be solved either on basis of the provisions of the sales contract or under the PRC Product Quality Law.

[352.] Id.

[353.] Id.

[354.] Insurance law of PRC [online] available at https://www.ilo.org/dyn/natlex/docs/Electronic/85811/96281/F307694451/CHN85811.pdf (last visited March 25, 2021).

PRC Bureau of State Quality Control, Inspection, and Quarantine had on 29 December 2012 promulgated the Automobile Guarantee Provisions which contains various provisions regarding Liability for Household Automotive Products. As per this, any seller who sells consumer automobiles will mandatorily assume three guarantee liabilities towards their consumers[355]. As per this regulation, passenger cars purchased and utilised by individuals for personal purposes are considered home automotive products[356]. These vehicles should be complemented by a maintenance handbook, a user's guide and a voucher with three guarantees, and a product inspection certificate in Chinese or another appropriate certification. The three guarantee liabilities[357] are guarantee liability for repair, replacement, and return. Post the sellers assuming liabilities towards their consumers on account of these guarantees preferably through a contractual arrangement with the manufacturers/dealers, the sellers can ask to be compensated by manufacturers/dealers for any liability attributable to the manufacturers/dealers.

As per PRC laws, the minimum warranty for a household automobile should be three years or 60,000 kilometres. The three warranties mentioned above should be valid for two years or 50,000 kilometres. Both these periods start from the date of the invoice provided to the consumer[358]. According to the PRC government's Administrative Measures on Automotive Sales, which were announced on April 5, 2017, auto suppliers (including automakers) are required to guarantee and ensure the supply of car components and related after-sales support for a minimum period of 10 years after the public announcement regarding production stoppage or sale of relevant vehicle models[359]. There is one more law in PRC for recalling defective vehicles. All manufacturers are obliged to recall sold vehicles if any defect is later found in them which needs attention at the manufacturers level. This liability is not contractual and is a statutory obligation imposed by law[360].

---

[355] Automobile Guarantee regulations for Automobiles used by Consumers [online] available at https://cms.law/en/chn/publication/automobile-guarantee-regulations-for-automobiles-used-by-consumers-have-finally-been-promulgated (last visited March 25, 2021).

[356] Id.

[357] Id.

[358] Id.

[359] Id.

[360] Autonomous vehicles law and regulation in China, CMS Expert Guides, https://cms.law/en/int/expert-guides/cms-expert-guide-to-autonomous-vehicles-avs/china (last visited April 29, 2021).

3.4.2 *Indonesia*

The Association of Indonesian Automotive Manufacturers (Gaikindo) hosted an exhibition in August 2018 to display innovations in the automobile sector. This exhibition facilitated manufacturers and innovators to showcase case latest AV technology from Indonesia to the rest of the world. The biggest problem that Indonesia faces is the traffic congestion on the roads of Jakarta which are prohibitive to adopting AV technology[361]. Also, the AV manufacturers in the USA have not considered or researched the effectiveness of AVs on the roads of non-Western countries.

Indonesia's government has declared the shifting of its capital city from Jakarta to the present forest area in North Penajam Paser. This is expected to be completed by 2024 for Rp 466 trillion (US \$32.7 billion)[362]. The Indonesian President has regularly been stating his desire for a smart and green new capital and hence his choice of AVs and EVs in the new capital city to make it the first capital city in the world to ever adopt AVs and EVs. Even while this is igniting a new debate on topics like the design of a new palace and the likelihood of obtaining international money during the pandemic, it appears that this will happen soon. The smart capital city is expected to use a lot of AI technologies[363]. As per the Indonesian President, the aim is to build the new capital city in such a manner that everything is efficient and cheap.

Indonesia has no separate regulations for AVs and for the use of biometrics which would be integrated into AVs. Non-availability of any such regulation cannot be construed as non-recognition of biometrics in Indonesia. As early as 2009, the Indonesian government had integrated biometrics into their brand-new electronic resident card, known as the eKTP. The basis for this was the Residential Administration Law, also known as Law No. 24 of 2013 on the Amendment of Law No. 23 of 2006[364]. As per said law, the biometrics of Indonesian residents qualify as personal data. These must be stored by the

---

[361.] Charlotte Trueman, the state of autonomous vehicles in Southeast Asia CIO (2019), https://www.cio.com/article/3309917/the-state-of-autonomous-vehicles-in-southeast-asia.html (last visited April 29, 2021).

[362.] The Jakarta Post, Jokowi wants all vehicles in new capital city to be autonomous, electric The Jakarta Post, https://www.thejakartapost.com/news/2020/01/15/jokowi-wants-all-vehicles-in-new-capital-city-to-be-autonomous-electric.html (last visited April 30, 2021).

[363.] Id.

[364.] Law 23 of 2006 on Population Administration [online] available at https://www.refworld.org/docid/54eeefce4.html (last visited March 26, 2021).

Government of Indonesia. However, this law does not explicitly call out if the law is applicable for other uses of biometrics, e.g., use in AVs and phones.

As regards data privacy Regulation No. 20 of 2016 (DP Regulation) from the Minister of Communication and Informatics (MOCI) is equally applicable to data collected by AVs or AV developers since Indonesia at present does not have a personal data protection law covering a broader range of subjects. According to the Data Protection (DP) Regulation, MOCI must first approve any overseas transfers of personal data[365]. The consent must be provided in writing and may be delivered physically or electronically, according to MOCI Regulation 20. Hence under normal circumstances, AV developers would be required to store data in Indonesia. Also, the present law is silent on whether the onshore data centre must be owned by the AV developers, or could be outsourced. Currently public and non-public service providers in Indonesia outsource data storage to third-party service providers. With proper permission, AV developers can store the biometrics data offshore. However, before such storing, as per MOCI 20/2016 AV developer must give notice of such offshore transfer to the data owners and seek their consent[366] before transferring such data. There is no restriction to having the consent in a bilingual format, however, it must be in the Indonesian language. In any instance, the DP Regulations do not mandate that the consent be contained in a separate stand-alone document. In line with the relevant laws and regulations, parental or guardian consent may be given if the data subject is a minor. The guardian must be someone who has a legal duty to look after the minor, whereas the parents must be either the biological mother or father[367]. In case of a breach, AV developers must notify the data owners in writing within 14 days after the incident.

As regards AV Patents, the process that one would fall back on would be the existing Patent laws of Indonesia. As per the principle of first registration under Indonesian Patent[368] laws, it is mandatory to register all patents in the Indonesian Patent Registry. A potential issue exists while

---

[365] Data Protection in Indonesia: Processing requirements [online] available at https://www.ssek.com/blog/data-protection-in-indonesia-processing-requirements (last visited March 27, 2022).

[366] Id.

[367] Id.

[368] Autonomous vehicles in Indonesia, Inside Tech Law | Global law firm | Norton Rose Fulbright, https://www.insidetechlaw.com/autonomous-vehicles/08_indonesia#:~:text=Indonesia%20currently%20does%20not%20have,Informatics%20(MOCI)%20Regulation%20No. (Last visited April 30, 2021).

acquiring a patent in Indonesia. Law No. 13 of 2016 on Patents (Patent Law) stipulates that to hold an Indonesian patent, the patent holder must produce or process the goods there[369]. The Paris Convention for the Protection of Industrial Property has been ratified by Indonesia, so the owner of the patent in the country of origin will reserve priority rights for the patent to be registered first in Indonesia (subject to the country of origin being a party to the Paris Convention or not). Post this Indonesia would cite the invention's date of patent registration in its nation of origin[370]. The local manufacturing law is mainly to ensure that the patent holder supports the transfer of technology, encourages investment, and helps in increasing jobs. Many stakeholders including governments from other countries have protested this regime.

3.4.3 Japan

Another major automobile manufacturing nation in APJC is Japan. This country is famous for its innovation and the kind of infrastructure it provides for research and development in the automobile space. Japan has been continuously seeing and forecasting the need for AVs and feels that the AV sector needs to be deregulated and promoted. Since Japan is strong on innovation, it is not surprising to know that it occupies the fifth position in the world when it comes to patents on AV[371]. Japan is also number three in the world when it comes to infrastructure and holds the best ratings for both 4G coverage and road logistics infrastructure. As regards policy, legislation, and customer acceptance it occupies a middle slot and is at the lower end of the stack for online ride-hailing.

One-fourth of Japan's population is older than sixty-five and as per many researchers including the famous KPMG group, which is one of the big fours in the consulting area, this age bracket will influence the use of AVs in Japan to a great extent. AVs will help provide mobility to this age bracket, particularly in rural Japan. There are many towns and villages in Japan where

---

369. Id.

370. Id.

371. Top five best equipped countries to support AVs [online] available at https://electrek.co/2022/03/04/the-top-five-best-equipped-countries-to-support-autonomous-vehicles-whos-leading-the-self-driving-revolution/ (last visited March 31, 2021).

there are no doctors, and millions of elderly people struggle to shop plus have had to return their driving licenses due to health issues. In 2018 many AV manufacturers in Japan carried out proof of concept (POC) exercises in Japan especially in the public space like taxis/buses that too within well-defined areas which also included airports.

The Japanese government is fully aware that to bring AVs to Japan they will have to ensure the existence of strong regulations and take care of technical problems which can crop up due to the regular snowfall, earthquakes, and tsunamis. Another major problem to be tackled in Japan is the ageing infrastructure i.e., roads, tunnels, and their high maintenance costs. Japan also believes that the use of AVs will accelerate in the coming years, and they do have a great political will to support this movement.

As of now Japan has plans to use SAE level 3 AVs[372] in the Olympic games 2020 (which was moved to 2021) and thereafter roll out SAE level 4 vehicles[373] for testing and the final commercial offer by 2025. There are also plans to do tests in the marine sector or remote and sparsely populated areas. The Japanese government's primary goal is to implement level 3 first and use it as an advantage to touch level 4. In level 4 as we all know system takes over driving under certain circumstances[374]. The expectation is to see level 4 AVs on Japanese highways by 2025. There are some level 3's where the car slows down if the driver does not react and they are considered half level 4.

As regards the Japanese Motor Vehicle Act, to date, the said act was designed to only handle AVs up to SAE level 3. With the latest revisions to the Act i.e., Law No 14 and 20 enacted in 2019 the act now has adequate provisions to handle level 4 SAE AVs also. The important provisions of the revised act are[375] that the AVs must adhere to safety regulations (revised Vehicle Act Section 41, clause 1, 20–2). Ministry of Land, Infrastructure, Transport,

---

[372] Outline of systematic preparations related to autonomous driving (2018), https://www.kantei.go.jp/jp/singi/it2/kettei/pdf/20180413/auto_drive.pdf, p.6 (last visited April 30, 2021).

[373] ITS 2019 (2019), https://www.kantei.go.jp/jp/singi/it2/kettei/pdf/20190607/siryou9.pdf, p.21 (last visited April 30, 2021).

[374] Staff Writer, Back-seat driver: How Honda stole the lead in autonomous cars Nikkei Asia (2021), https://asia.nikkei.com/Spotlight/The-Big-Story/Back-seat-driver-How-Honda-stole-the-lead-in-autonomous-cars (last visited April 30, 2021).

[375] Legal regulation of autonomous driving technology: Current conditions and issues in Japan [online] available at https://www.sciencedirect.com/science/article/pii/S0386111219301591 (last visited, March 31, 2021).

and Tourism (MLIT) shall prescribe the circumstances in which each form of AV is to be driven based on its performance (revised Vehicle Act Section 41, clause 2)[376]. Level 3 SAE AVs must only operate inside the predetermined operational domain (ODD), which includes the driving environment conditions like roads with or without lanes, geographic conditions like mountain areas, natural environment conditions like night/daytime, and other prescribed conditions to ensure safety[377].

As per the earlier Japanese Motor Vehicle Act, disassembly and maintenance can be done only with the permission of the local transportation bureau's director. This has now been modified to include a concept called specified maintenance which also encompasses upkeep required to achieve autonomous driving (revised Vehicle Act Section 49, clause 2)[378]. Removing and repairing the braking mechanism has been thought to be a traditional example of conventional car disassembly and maintenance. In AVs, there are many other parts which control AV driving like Cameras, Radars and Lidars. As per the revised act, since installation and adjustment of cameras and radar devices are crucial for the safe driving of an AV, these activities can now be done only after obtaining specific approval.

AV manufacturers are now required to provide all technical information about their vehicles to technicians who are repairing or servicing their vehicles (revised Vehicle Act Section 57, clause 2–1). Failure to do so is a criminal act under the revised Vehicle Act Section 109, clause 8 with a maximum penalty of 500,000 yen fine[379]. Another new clause has been added to seek permission for specified modification of AVs through telecom lines (revised Vehicle Act Section 99, clause 3–1-1)[380]. According to updated Vehicle Act Section 100, adequate administrative measures would be implemented to guarantee that the conditions for the specified permit are satisfied, and in the event of any violations, fines might be imposed under revised Vehicle Act Section 109, clauses 14, 15.

With the passing of the two revised laws, there has been a great amount of reduction in legal obstacles for AVs to ply in Japan[381]. The Traffic Act will

376. Id.
377. Id.
378. Id.
379. Id.
380. Id.
381. Id.

be amended to take into account the prospect of AVs with level 3 driving technology operating on public roads. Although level 4 autonomous driving technology is anticipated to be used on public roads after 2020, several aspects of its development and associated issues up to that point cannot currently be foreseen. Accordingly, these revisions are not targeted directly at level 4 vehicles. However, many issues still have to be resolved and the most important of them is the liability of the AV in case the AV kills say one person to possibly save say five lives. As per penal laws of Japan, the AV owner is still guilty of killing one person but this is a subject that needs a law with a broader vision.

3.4.4 *Malaysia*

A Malaysian research and development (R&D) business by the name of REKA set out in 2016 on a mission to establish Malaysia as a global AV powerhouse. As a part of its research REKA created a chip to create its self-driving technology for AVs to be inserted along with sensors, GPS and a video cam[382]. The Proton Wira and Proton Perdana automobiles now have off-the-shelf sensors and cameras thanks to Reka's crew of 20 individuals. The components of the car were connected via a central processor called the RIG to run it (REKA Innovation Gear). As a part of the test drive an AV with no driver but only two passengers sitting in the rear seats plied successfully from Kuala Lumpur to Melaka. With the success of the pilot project, the Malaysian government has provided support to the project through an R&D program to be continued till 2025.

The Association of South East Asia Nations (ASEAN) response to Uber is Grab the popular ride-sharing app in Malaysia. Grab became the largest ride-sharing app in the area, which includes Indonesia, Philippines, Vietnam, Thailand, Myanmar, and Cambodia, after it bought Uber's Southeast Asian operations in March 2018. As of date Grab has partnered with a Singapore-based company called NuTonomy for developing and testing AV technology. Grab is quite optimistic that their AV taxis would be on Southeast Asia roads by 2022.

---

[382] Abu Kassim et al., *Is Malaysia* Ready *to Adopt Autonomous Vehicles?*, 3 Journal of Society of Automotive Engineers Malaysia 84–88 (2019).

There are gaps in the policies of the Malaysian government regarding AVs. As of date, Malaysia has no proper stand-alone laws or policies concerning AVs, which is very much essential to ensure the adoption of AVs and their safe deployment. The Malaysian government is open to adopting new technologies but without a proper law in place, the adoption may not be meaningful. Apart from enacting appropriate laws for AVs, lawmakers must also work with the AV industry to ensure its development and growth[383].

3.4.5 Philippines

There is now a programme in place in the Philippines to construct smart cities. One such smart city is called Clark Global City and is expected to act as a perfect testing ground for AVs in the Philippines. The various stakeholders of this smart city project are keenly ensuring that the project takes care of all aspects required to ensure AV development and also provides the impetus for AV-related start-ups to launch and thrive[384]. The Bases Conversion and Development Authority (BCDA) is collaborating with the US-based COAST Autonomous mobility start-up to conduct free government-funded low-speed autonomous vehicle pilot testing within New Clark City. At the South East Asian Games in 2019, the competitors were transported using autonomous or self-driving vehicles[385]. As of date, there are no AV-specific laws in the Philippines to address liability and other related aspects. The Filipino Insurance Commission has cautioned the government to carry out a thorough reform of several legislations to bring about a coherent environment for AVs to operate in the Philippines because the insurance business there is concerned about the potential consequences if AV technology is deployed[386].

[383.] Id.

[384.] BCDA to pilot autonomous vehicles in New Clark City for SEA Games [online] available at https://bcda.gov.ph/news/bcda-pilot-autonomous-vehicles-new-clark-city-sea-games (last visited March 31, 2021).

[385.] Id.

[386.] Id.

3.4.6 *Singapore*

The Singapore government is quite positive in their plans to make the city an AV development centre. The government has plans to support plying of driverless buses in the city and also has simulated urban test beds. In November 2017, Singapore opened up a Centre of Excellence for Testing and Research of AVs at Nanyang Technological University (CETRAN). This CETRAN has a realistic test bed with bus stops, traffic lights, huge buildings, mountains, and an artificial rain generator[387]. CETRAN is also in collaboration with The Netherlands Organisation for Applied Scientific Research (TNO) to help accelerate the safe introduction of AVs in the Netherlands and Singapore. As per the announcement made by the Singapore government, from 2022, driverless buses and shuttles are anticipated in Punggol, Tengah, and the Jurong Innovation District for off-peak and on-demand travel.

The Singapore government is very proactive when it comes to discussions on the future of mobility, especially in the AV sector. As per KPMG reports Singapore stands at 15th position out of 25 when it comes to technology and innovation. Visionaries feel that since conventional automobile manufacturing is not there in Singapore, this could be their strength for AV manufacturing as AVs use a substantially different set of components[388]. By declaring that they would establish a new facility in Singapore to produce electric vehicles in 2018, Dyson confirmed this[389]. Since the taxes for private cars is quite heavy in Singapore, that's another driving factor to promote AVs.

As per the Road Traffic (Autonomous Motor Vehicles) Rules 2017 (2017 Rules), the trial of AVs in Singapore is permitted only if one has permission from the Land Transport Authority of Singapore (**LTA**). The LTA is empowered to grant permission only in a specific area for the trial along with other conditions[390]. An autonomous motor vehicle is entirely or primarily equipped with an autonomous system, as defined by the Road Traffic Act (Cap 276)

[387.] Automotive Essentials [online] available at https://www.tuvsud.com/en/e-ssentials-newsletter/automotive-essentials/e-ssentials-2-2017/project-cetran-preparation-for-the-automated-driving-revolution (last visited April 1, 2021)

[388.] 2020 Autonomous Vehicles Readiness Index (2020), https://assets.kpmg/content/dam/kpmg/it/pdf/2020/07/2020-autonomous-vehicles-readiness-index.pdf (last visited April 30, 2021).

[389.] Dyson picks Singapore for its new Electric car factory [online] available at https://www.cnbc.com/2018/10/23/dyson-picks-singapore-for-its-new-electric-car-factory.html (last visited April 3, 2021)

[390.] Road traffic Autonomous Motor Vehicles Rules 2017 [online] available at https://sso.agc.gov.sg/SL/RTA1961-S464-2017?DocDate=20170823 (last visited March 25, 2021).

(RTA) (also commonly known as a driverless vehicle)[391]. This definition has no mention of SAE Levels. Persons authorised by LTA have to ensure that the AV is properly maintained and that it causes no harm or damage to any person or property. The authorised persons must inform LTA regarding any incident involving malfunction of the AV or any accident by the AV causing physical hurt, death, or property damage to someone. Any breach of the above responsibilities can result in conviction and imposition of associated liabilities.

The AV owner is required under the 2017 Rules to mandatorily have a liability insurance policy for the vehicle and if the owner is not able to do that then a security deposit must be placed with the LTA to compensate for any payments due on account of any wrongful death, physical harm, or property damage[392]. The 2017 Rules also prescribe submission of certain supporting documents while seeking permission from LTA, including details of the AV which will be used in the trial with an undertaking that the AV is safe for trials[393]. The LTA before approval might require that a properly trained safety operator or driver must be seated in the AV at all times to monitor its operations and take over control if required. The driver's or operator's liability would be decided as per current laws[394].

The 2017 Rules do not specifically prescribe information on the manufacturer's responsibility if an AV is operated automatically. However, existing manufacturer's obligations will be applicable[395]. The obligations would include a duty to report non-conformities or defects under the Sale of Goods Act (Cap. 393, Rev Ed 1999) and Unfair Contract Terms Act (Cap. 396, Rev Ed 1994) to the Registrar of Vehicles and the vehicle owners[396].

Under the RTA framework, AV trials can only be carried out post procurement of liability insurance for damage to life/property till the duration of the trial or placing a security deposit of an amount not less than S $1.5 million for the trial duration with LTA[397]. The courts would make the current general product liability or consumer protection legislation relevant for warranties.

[391.] Id.

[392.] Id.

[393.] Id.

[394.] Autonomous vehicles law and regulation in Singapore, CMS, https://cms.law/en/int/expert-guides/cms-expert-guide-to-autonomous-vehicles-avs/singapore (last visited April 30, 2021).

[395.] Id.

[396.] Id.

[397.] Id.

3.4.7 South Korea

On December 10th, 2018 in South Korea, the K-City test facility, an experimental city setting outfitted with 5G technology to test AVs, launched. K-City is located in Hwaseong and was built by Korea's national telecom service provider Korea Telecom (KT) and the Korea Transportation Safety Authority. K-City's focus will be to test and commercialize AVs, especially Level 3 AVs which would also include sophisticated features but still require a safety driver who will take over driving duties as needed[398]. Five key testing locations may be found in the K-City: a highway, a downtown road, a suburban street, a parking lot, and community facilities.

Post opening of K-City, as a part of its innovation strategy, KT managed to successfully test an AV bus at South Korea's Incheon airport. This bus successfully changed lanes and also stopped at traffic lights. The South Korean Ministry of Land, Infrastructure and Transport has also ventured into a project to create detailed maps for AVs. This would be a part and parcel of Cooperative Intelligence Transport Systems (C-ITS), which is a project developed for enhancing the safety of AVs. Samsung another technology giant in South Korea released a platform for AVs that combines hardware and software. This was possible after Samsung's US $8 billion purchase of a US-connected car company, along with which Samsung also got control of the US $300 million automotive innovation fund[399].

Over 30 countries utilise the Autonomous Vehicles Readiness Index (AVRI), a global methodology, to assess the readiness of AVs. AVRI is a comprehensive index which measures 28 individual scores from a variety of sources to translate them into a single score. In most countries, AVRI is generally used by public sector organizations which are responsible for transport and infrastructure. Other public and private sector businesses that work in or use the transportation industry also use AVRI[400].

As regards, AVRI South Korea has a great score on several AVRI measures. Some of these like AV pilots, EV charging stations and 4G coverage are funded by the government. As regards the legislative processes, South Korea is poorly

---

[398.] South Korea builds K-City for AV testing [online] available at https://en.yna.co.kr/view/AEN20181210008800320 (last visited March 25, 2021).

[399.] Samsung is buying Harman in an $8billion bet on connected cars [online] available at https://www.theverge.com/2016/11/14/13620812/samsung-cars-harman-acquisition-8-billion (last visited March 23, 2021)

[400.] 2020 Autonomous Vehicles Readiness Index, Supra note 403, at 109.

rated under AVRI. In comparison with other countries which are pioneers in the AV sector, South Korean legislation only permits the driving of level 3 AVs due to its slower legislation process. Korea is now 13th in the AVRI index after it was overtaken by Norway and Finland and Japan outpacing it in the technology and innovation pillar[401].

3.4.8 *Thailand*

For a long time, Thailand has been considered one of the prominent automation production hubs of Southeast Asia. The auto industry in Thailand helps Thailand generate 12% of its gross domestic product by making around 2.2 million vehicles in a year on average. To keep up with international trends, the Thai government is also attempting to increase investment in EVs and AVs[402]. The Thai government is doing this by offering incentives to Electric Vehicle (EV) manufacturers by reducing sales taxes on AV/EV cars. The Thai government is also supportive of building AV vehicles on EV platforms.

As regards the adoption of AV technology Thailand has been lagging behind most countries in Southeast Asia. Thailand does not figure in the top 25 countries where Singapore is in second place, Japan in the tenth and South Korea in thirteenth followed by China in the twentieth. Amongst the many roadblocks which are hindering the introduction of AVs in Thailand[403] current Thai regulations are not compatible with AVs and hence the test AVs cannot be even taken out on the public road, despite Thailand being one of the champions of AV technology in Southeast Asia. When one compares Thailand with Singapore, it is visible that the quality of infrastructure in Thailand is poor compared to Singapore. In many parts of Thailand, roads are not laid down as per approved plans and this can result in navigational inaccuracy. Traffic lights and road signs throughout Thailand are improper and this can again mislead AVs. Finally, as of date, there is no market for driverless AVs in Thailand. Despite it being more populated than Singapore and Vietnam,

---

[401.] Id.

[402.] Tuk-tuk in Thailand to get a makeover, government plans to transform commute with autonomous vehicles, The Economic Times, https://economictimes.indiatimes.com/magazines/panache/tuk-tuk-in-thailand-to-get-a-makeover-government-plans-to-transform-commute-with-autonomous-vehicles/articleshow/71389064.cms?from=mdr (last visited May 5, 2021).

[403.] Id.

Thailand is not considered a forerunner in adopting emerging technologies like AV.

Thailand is famous for its three-wheeler called tuk-tuk and a recent public-private partnership will ensure that we soon have AV tuk-tuks in place. Thai authorities, investor Siri Ventures, and start-up Airovr intend to conduct AV tuk-tuk testing inside a gated Bangkok neighbourhood[404]. The gated community has more than 10,000 people and a testing area of around 20 acres. There are around eight residential towers, a hospital, a school and other buildings in the testing area. To assist the tuk-tuk in collecting real-time data for additional testing and analysis, other vehicles such as automobiles, motorbikes, bicycles, and pedestrians will be operating on the streets. Post such real-time testing it will be possible to test and use AV technology in bigger public transports like minibuses. Plans are on to manufacture 15-seat minibuses named Shuttlepod which would be produced by the government in conjunction with a local automaker. As of date, there are driverless shuttles plying in New York's Brooklyn Navy Yard, between the new ferry dock and the entrance[405]. Similarly in Detroit, a start-up called May Mobility Inc. runs a fleet of six-seater robotaxis. These robotaxis transport the workers of May Mobility Inc from the parking place to the office[406].

Tuk-Tuk was chosen for AV experiments because it is more energy efficient than a car, has fewer parts, is less expensive, and is best suited for Thailand's hot and muggy climate[407]. The Airovr has a 3D mapping system on its roof akin to police sirens, and the vehicle has handlebars on the inside to help a tester to take control if necessary. However, the Airovr model is based on a minimalist design including a screen which displays the speed and balance of fuel/electricity. The Department of Land Transport and the Department of Highways in Thailand are working with other government agencies to formulate plans for Thai roads to facilitate the smooth adoption of AVs in Thailand. As of date the target for AVs to be on Thailand roads is 2024. Initially, it would be operators of public fleets and ride-hailing services.

[404.] Id.

[405.] Id.

[406.] Id.

[407.] Tuk-tuk in Thailand to get a makeover, government plans to transform commute with autonomous vehicles, THE ECONOMIC TIMES, https://economictimes.indiatimes.com/magazines/panache/tuk-tuk-in-thailand-to-get-a-makeover-government-plans-to-transform-commute-with-autonomous-vehicles/articleshow/71389064.cms?from=mdr (last visited May 5, 2021).

3.4.9 *Vietnam*

The automobile industry in Vietnam has been growing leaps and bounds in the last few years but is still way behind when we compare it with its neighbouring countries. According to the Vietnam Automobile Manufacturers Association, Vietnam sells about 270,000 cars annually. Even though the appetite for cars is increasing among the middle class in Vietnam, motorcycle sales are high in Vietnam approximately 3 million per year. Vietnam as a whole is more an automobile assembling than a manufacturing country[108]. Most automobile parts are imported.

The Vietnamese government has under the ASEAN Agreement on Trade in Goods, which governs trade in goods, come up with some tariff reduction schemes and these could open up the manufacturing market in automobiles. 2017 saw the introduction of the first AV (car) in Vietnam. This was made possible through software developed by a local company called FPT Software[409]. This company tested AVs by operating a test programme for driverless cars in Vietnam's technological parks @25 kilometres per hour and employed computer vision, radar, and GPS to monitor the environment and prevent accidents. As of date, there is no special law governing AVs in Vietnam and hence only existing motor vehicle, insurance and other consumer regulations will govern AVs till a special law is in place.

The Ministry of Transport and the Ministry of Industry and Trade regulate and supervise the automobile industry in Vietnam. The Ministry of Transport looks after technical and environmental standards and the Ministry of Industry, and Trade regulates imports of automobiles and automobile parts. Local motor vehicle assemblers who assemble completely knocked down (CKD) vehicles are required to comply with environmental and technical standards prescribed by the Vietnam Ministry of Transport and obtain relevant approvals as per Circular No. 30/2011/TT-BGTVT of the Ministry of Transport[410]. The final products can be subjected to inspection by the Ministry of Transport.

Completely built-up (CBU) vehicles are vehicles which are fully imported in assembled state. These CBU vehicles are required to complete

---

[408.] Jeff Oslon & Huong Phan, Automotive in Vietnam Lexology (2019), https://www.lexology.com/library/detail.aspx?g=94f93be9-e1be-4d62-a44c-8c6995229650 (last visited May 5, 2021).

[409.] Việt Nam's first autonomous vehicle debuts, vietnamnews.vn, https://vietnamnews.vn/economy/914897/viet-nams-first-autonomous-vehicle-debuts.html (last visited May 10, 2021).

[410.] Ministry of Transport [online] available at https://vanbanphapluat.co/circular-no-30-2011-tt-bgtvt-on-the-environment-protection-and-technical-safety (last visited March 24, 2021).

a homologation process to be certified by the Vietnam Register. Exempted vehicles are required to obtain an exemption certificate[411]. As per Decree 116/2017/ND-CP effective from 1 January 2018, automobile importers need an import license from the Ministry of Industry and Trade which is only provided post rigorous quality inspection requirements and compliance with a lot of business requirements. All individual vehicles must be registered with the appropriate police department, and their owners are responsible for making sure the vehicles are properly insured[412].

The relevant vehicle-related product compliance safety and environmental regulations of Vietnam are outlined in Circular No. 30/2011/TT-BGTVT dated 15 April 2011 (governing the production of CKD vehicles), Circular No. 31/2011/TT-BGTVT dated 15 April 2011 (governing the import of CBU vehicles), and Circular No. 70/2015/TT-BGTVT dated 9 November 2015 (setting out duties applicable to car owners)[413]. Serious penalties, fines, and the loss of pertinent licences and certificates are possible outcomes of non-compliance.

As regards, product recalls, as per Circular No. 30/2011/TT-BGTVT, products failing to satisfy any applicable technical requirement, or which are likely to endanger humans or property due to such technical errors have to be recalled by the manufacturer within five days of detecting the error[414]. In such a case the manufacturer is also required to notify the Vietnam Register in writing with a proposed recall plan. Similar to this, new car emission regulations have been increased from Euro 2 to Euro 4 starting January 1, 2017, and from Euro 5 starting January 1, 2022, under Prime Minister's Decision No. 49/2011/QD-TTg. As regards product liability Vietnam's Civil Code and the Law on Consumer Rights Protection prescribe the right guidance to calculate the liability for damages.

---

[411] Id.

[412] Decree No 116/2017/ND-CP [online] available at https://english.luatvietnam.vn/decree-no-116-2017-nd-cp-dated-october-17-2017-of-the-government-on-requirements-for-manufacturing-assembly-and-import-of-automobiles-and-trade-in-117725-Doc1.html#:~:text=Decree%20No.,automobile%20warranty%20and%20maintenance%20services (last visited March 24, 2021).

[413] Ministry of Transport, Supra at 425, at 114.

[414] Id.

3.5 Existing laws, regulations, and policies on Autonomous Vehicles in ANZ

To boost AVs Australia now has a federal body for transport technologies. As regards scoring on regulations supporting AVs is concerned, Australia is amongst the top scorers as they have had broad changes in their motor vehicle and other connected policies and regulations. On infrastructure scoring also Australia has seen significant improvement. Despite the addition of five new countries and also two amongst those i.e., Norway and Finland are in the top six, Australia only dropped one spot in the overall 2019 rankings[415].

Australian government started the process of phased transformation of existing driving laws in 2016 to enable the complete use of AVs from 2020. To achieve this goal in October 2018 the Australian federal government set up the Office for Future Transport Technologies. This office was set up by investing 9.7 million Australian dollars (A $) (US $7.1 million) purely to bring together the states and government agencies for delivering future transport technologies safely and responsibly[416]. The recent MOU between the Australian Government and the US State of Michigan, in partnership with ITS Australia, that will facilitate future opportunities, is just one example of how this new Office offers an appropriate structure to support international collaboration on future vehicle technology. Towards this setup, South Australia, Victoria, New South Wales, and Queensland have all made significant investments. Net-Net all wings of the government are working in close cohesion to support AV deployment through the collaboration for trials, building of infrastructure and development of laws and regulations.

Australia has a road operator association known as Austroads, which is working on projects to support AV operations through machine-readable signage and road marking[417]. Austroads also works in the areas of open data specifically data received from road operators and does a lot of research especially looking into likely prospects for heavy AVs in regional and rural settings. Apart from Australian government agencies, Transurban a toll road operator in Australia has conducted trials of AVs along with six vehicle manufacturers. Australia is now EV friendly, and one can easily now go around Australia in an

---

[415.] 2020 Autonomous Vehicles Readiness Index, Supra note 403, at 109.

[416.] Office of Future Transport technologies launched [online] available at https://www.spatialsource.com.au/federal-office-of-future-transport-technologies-announced/ (last visited March 24, 2021).

[417.] Austroads [online] available at https://austroads.com.au/ (last visited April 4, 2021)

EV as there has been a significant installation of charging points for EVs across Australia. As per a publication brought out in October 2018 by Infrastructure Victoria (IV), an advisor to the Government, it is predicted that by 2046, there is a fair possibility that with AVs, annual greenhouse gas emissions would drop by 27 million tons[418]. IV has also predicted that this change will bring about a 91 percent increase in Victoria's Road network efficiency and is also expected to boost Australia's economic growth by around $A15 billion (US $10.8 billion) annually.

In addition to the above there are various influencers who have commented that despite the existing progress being made by Australia in the AV space, there still exists enough opportunities for AV integration in Australia. Some of these opportunities could be in areas of implementing recommendations towards infrastructure development and also addressing concerns and sentiments of consumers. EVs and AVs are the vehicles of the future and as they become more popular there is a growing demand on the government to understand new energy demand patterns and find innovative solutions as traditional gasoline usage eventually declines, to replace profits.

As a country, New Zealand has an excellent reputation to develop new technologies even though further refinement of these technologies happens elsewhere. This is because domestic manufacturing in New Zealand lacks scale. A company called Ohmio did trials of its driverless buses at Christchurch airport[419]. The airport's second round of experiments will not include the shuttle Navya, which is smaller than the Lift and has been tested on local roads since the launch. It was loaned to the airport by tech company HMI. Post trials Ohmio made a public announcement that it will be starting a joint venture with the city named Heshan in China and that most of its manufacturing and research are expected to move to Heshan.

An area of Christchurch in New Zealand post the 2011 earthquake was categorized for redevelopment and has been identified and approved by the New Zealand government as a location for AV trial. This place would also

---

[418]. Advice on automated and zero emission vehicles [online] available at https://www.infrastructurevictoria. com.au/wp-content/uploads/2019/04/Advice-on-automated-and-zero-emissions-vehicles-infrastructure-accessible-version.docx (last visited, March 25, 2021).

[419]. Christchurch Airport buys new driverless electric shuttle [online] available at https://www.stuff.co.nz/technology/113672129/christchurch-airport-buys-new-driverless-electric-shuttle (last visited June 23, 2021).

be known as Mobility Lab[420]. With Mobile Responder, officers can access real-time vital data and incident communication from Intergraph's computer-aided dispatch system, I/CAD, via their smart devices over Vodafone's high-speed network while on the go. With the mobile solution, officers can now make smarter decisions faster in critical situations. New Zealand is also testing autonomous flying taxis through a company called Cora.

Since New Zealand's domestic market is small and most of the technology companies operating out of New Zealand are foreign-owned, no automobiles are produced in New Zealand, nor is any type of mainstream research done there. The relatively small population in New Zealand is a challenging factor for the adoption of AVs. In addition, the patchy networks, and the mixture of vehicle fleets from Europe (new cars) and Asia (used cars) are another set of challenging factors which obstruct the smooth adoption of AVs in New Zealand. Despite the above handicaps, New Zealand is well placed in policy and legislation areas which can help the growth of AV. The best part of New Zealand lawmakers[421] and regulators is that they all have a can-do attitude and are strongly focused to find solutions and remove barriers if any. New Zealand Police enjoy huge productivity gains from the mobility programme, effectively saving around 30 minutes per officer per shift, adding to a massive 520,000 hours each year, and most significantly, putting 345 more police officers on the streets. While New Zealand may not step into full-fledged AV manufacturing, there is a strong possibility of New Zealand developing AI and AV controls required for AVs. Such a development can help New Zealand focus on specific areas like agriculture. Apart from agriculture, these AV controls can also be used in drones and airports.

3.6 Regulations of Self-Driving Cars or Autonomous Vehicles in the Rest of Europe

The Bulgaria Road Traffic Act 1999 (**Road Traffic Act**)[422] does not permit the testing of automated vehicles on roadways in Bulgaria. As per the Road

---

[420] Inside New Zealand Police's Mobility Innovation Lab and Experience Centre [online] available at https://blog.hexagonsafetyinfrastructure.com/inside-new-zealand-polices-mobility-innovation-lab-and-experience-centre/ (last visited June 23, 2021).

[421] Id.

[422] Road Traffic Act [online] available at http://www.lawyersbulgaria.org/content/2011/603/road-traffic-act/ (last visited March 21, 2021).

Traffic Act, all vehicles on the road must have a driver. Hence no consumer is permitted to use an AV on Bulgarian roads. Since AVs are not permitted in Bulgaria the Road Traffic Act is silent on liability for AV accidents. For conventional cars, the driver and sometimes the owner can be held liable for damages. The updated Bulgarian Insurance Code 2016 (Insurance Code) additionally requires third-party liability insurance, which must be carried by all motor vehicle owners, users, and holders as well as by anybody who operates the vehicle for legitimate purposes[423]. As per the Bulgarian Consumer Protection Act 2005, the manufacturer of a product, its representatives and in certain cases, the seller has to bear liability to ensure the safety and compliance of the product.

Estonia is a country in northern Europe which has more than 1500 islands with diverse terrains, rocky beaches, forests and lakes. Once upon a time, Estonia was part of the erstwhile Soviet Union and its capital city is Tallin which is famous for its 314m high TV tower with an observation deck. Pre-2017 Estonia had allowed the testing of AVs on the roads in the country[424]. The chance to promote Estonia as a nation accepting of new ideas is provided by the presidency. According to Valdek Laur, who is in charge of creating a digital representation of Estonia's EU presidency, self-driving cars used for public transit might significantly improve road safety and contribute to environmental conservation. Post-2017 Estonia permitted the usage of AVs by the public with a condition that the AVs must have a safety driver inside who must be capable of assuming control of the vehicle when necessary.

In May 2018 Finland introduced a driverless Robobus with a nondriving human operator. Finland plans to introduce commercial driverless buses in 2021[425]. To support this a Finnish company called Sensible is developing an AV minibus, partnering with Japanese designer Muji, which will also work in harsh Finnish winters. Finland has supportive regulations for AV. The Finnish Traffic Safety Authority does not withhold permits for testing AVs and does

---

[423] Insurance Code [online] available at https://www.abz.bg/public/uploads/files/code_en.pdf (last visited March 21, 2021)

[424] Self-Driving Min Buses set to roam in the streets of Tallin in 2017 [online] available at https://e-estonia.com/self-driving-minibuses-set-to-roam-the-streets-of-tallinn-in-2017/ (last visited March 21, 2021).

[425] New Autonomous Shuttle Bus unveiled by Finland's 4 OY and Moove GMBH [online] available at https://www.autofutures.tv/2021/12/02/new-autonomous-shuttle-bus-unveiled-by-finlands-sensible-4-oy-and-moove-gmbh/ (last visited March 21, 2021).

allow testing on all Finish roads. Finland's government has recently passed two laws to enable AVs[426]. The country now permits AV Taxies with a safety driver who can control the vehicle remotely. The amended Road Traffic Act looks into integrating data relating to the vehicle location, traffic lights, road signs and other control mechanisms which would be used by AV operators. Since white lines on roads are easier for an AV to detect than yellow lines, the above Act has approved the repainting of yellow lines to white.

Hungary is another European country which has initiated baby steps towards the usage of AVs. In 2017, Hungarian amended two old ministerial decrees i.e., 5/1990 and 6/1990 to permit the testing of AVs for experimental purposes. Post this amendment road testing of AVs was legalised in Hungary[427] subject to specific prior permission from the Ministry of Traffic. The Ministry can refuse permission if the proposed testing does not comply with certain legal and technical requirements. Laws in Hungary don't use the words self-driving vehicle or AV.[428] Neither the terms "autonomous vehicle" nor "self-driving car" is used in Hungarian regulations. Hungarian lawmakers instead proposed a more inclusive phrase. An experimental autonomous vehicle is one that: (a) aspires to develop partially or automated operation; and (b) has a trained test driver who, depending on the level of automation of the vehicle, can exert manual control when it is necessary for circumstances that endanger traffic safety[429]. As of date consumers are not permitted to drive AV on the roads in Hungary. Annexe 18 of the said ministerial decree 6/1990 classifies AVs as per internationally recognised classification and Annex 17 contains detailed notes on the operational and technical aspects of AVs.[430] Since there are no specific rules concerning liability on account of accidents caused by AVs, the general liability law will determine the liability of the owner. Such liability can also include criminal liability.

As regards product liability, the manufacturer and/or the AV software developer would be subject to existing general strict liability rules for damages caused by the AV in "autonomous mode". The only exception would be if

---

[426] 2019 Autonomous Vehicle Readiness Index Supra note 277, at 71.

[427] Towards a European Regulation of Autonomous Vehicles - EU Perspectives and the German Model, https://www.researchgate.net/publication/333575970_Towards_a_European_Regulation_of_Autonomous_Vehicles_-_EU_Perspectives_and_the_German_Model (last visited March 2, 2021).

[428] Decree No 5/1990, Article §2 (3b), point b, Hungary.

[429] Id.

[430] Decree No. 6/1990, Annex 17 and 18, Hungary.

it is proved that the damage was unavoidable. In any case, liability will only arise when the damages are caused by a faulty part or solution applied by the developer. There is no explicit obligatory AV insurance requirement in Hungary and hence the general mandatory requirement that applies to any vehicle will apply to AVs too. Hence whenever AVs are tested on public roads, they must have in place appropriate insurance.

In February 2018, Italy passed its first law permitting the testing of AVs[431]. The law allowed testing only on specific roads. Further testing has to be authorised by the Appropriate Authority which is the Ministry of Infrastructure and Transport. Before such authorization, there is a mandatory requirement to have the AV insured for any related civil liability, up to a maximum of four times the legal minimum for a similar vehicle without autonomous driving. The use of the vehicle and the fact that it is driven automatically on public roads must be mentioned in the insurance policy. Also, the law specified that all AV testing must mandatorily have a safety driver in the vehicle, who can take control of the AV when required. AV manufacturers and research centres must certify that the AV is ready in all aspects for road testing in the place where it is to be tested. This certification must be made after considering all supporting infrastructure including but not limited to painted lines, road signs and traffic signals.

As of right now, consumers in Italy are not allowed to utilise AVs (SAE Levels 3-5) on public roads. The existing general liability regime which applies to all automobiles plying on Italian roads applies to AVs also, since there are no separate liability rules specific to AVs[432]. In case of any liability issues under normal circumstances, unless the owner can demonstrate that the car was driven against his will, the owner and driver would be equally and severally liable. As regards the manufacturer, the general liability rules would apply for defects in the vehicle, including the mandatory two-year consumer warranty for defects.

As a country, Israel is also one of the greatest AV innovators. However, it lacks domestic manufacturing resulting in poor local deployment. Daimler, the world-famous car manufacturer has invested in Israeli smartphone data analytics company Anagog. In parallel, there have been commitments from

---

[431.] Autonomous Vehicles Law and Regulations in Italy [online] available at https://cms.law/en/int/expert-guides/cms-expert-guide-to-autonomous-vehicles-avs/italy (last visited March 2, 2021)
[432.] Id.

many other popular vehicle manufacturers to locate R&D facilities in Israel. Most of Israel's businesses are export-oriented. Since Israel is a small country with fewer journey distances introducing AVs will not be a cakewalk. The coexistence of autonomous and human-driven vehicles on Israel's congested metropolitan highways would be a difficult job. Israel is in the process of coming out with a proper legal regime for the safe deployment of AVs. Israeli Ministry of Transport and Road Safety (MOT) is working on R&D for AVs and in addition is also focusing on Apps to provide Mobility as a Service (MaaS)[433]. The Israeli Ministry of Transport and Road Safety (MOT) is funding the development and testing of AVs while concentrating on MaaS and public transportation applications.

To help government organisations to adopt flexible and dynamic regulations to promote rapid technological changes, the Centre for the Fourth Industrial Revolution (C4IR) Israel was established[434]. The Israeli Innovation Authority, a focal point for fostering innovative ideas and serving the requirements of the Israeli Hi-Tech sector, has been given a mandate by the Government of Israel to act as the hosting organisation of the Israeli C4IR affiliate centre. C4IR and MOT work jointly to improve Israel's regulatory landscape to establish AV policy for Israel and aid in the easy transition to a shared, electrified, and automated mobility ecosystem. Thereafter MOT and C4IR will study the AV regulations and policies of Australia, Singapore, the United Kingdom, California, and Arizona to come up with the best legal regime for AVs in Israel.

As per Sovereign Order n° 7.489 of 27 May 2019 concerning plying of AVs on public roads, AVs are allowed to ply on the roads in Monaco for testing purposes with either full or partial delegation of driving on a public road. However, before conducting such testing permission from the Minister of State is to be obtained by the manufacturer following the procedure set out in Ministerial Order No. 2019-512[435]. As per the 2019 order, the application for permission must mandatorily specify the level of automation. As of date, Monaco does not have any legal regime which permits the use of AVs by

---

[433.] Autonomous Vehicle (AV) Policy Framework, Part I: Cataloging Selected National and State Policy Efforts to Drive Safe AV Development, https://innovationisrael.org.il/sites/default/files/Autonomous%20 Vehicle%20Policy%20Framework-.pdf (last visited March 12, 2021).

[434.] Id.

[435.] Autonomous Vehicles Law and Regulations in Monaco [online] available at https://cms.law/en/int/ expert-guides/cms-expert-guide-to-autonomous-vehicles-avs/monaco (last visited March 2, 2021).

consumers on public roads. Monaco is yet to enact laws to handle liability on account of an accident caused by an AV. Hence, in case of an accident by an AV, the general law on tortious liability would be applicable[436]. Article 1229 of the Civil Code, which states the following, governs the tort liability system in Monegasque:

"Any act of man that causes damage to another person obliges the person through whose fault it occurred to make reparation for it"[437] (Article 1229, Monaco Civil Code).

As regards insurance for AVs there is no specific law in Monaco. However as per Ministerial Order No. 2019-512 AVs being tested need specific Insurance. As per the existing Ordinance, No. 2.617 of 23/08/1961 compulsory insurance is required for all motor vehicles[438]. Going by this law all AVs irrespective of their SAE level would be subject to the above-mentioned compulsory insurance obligation. For liability due to accidents caused by AVs, there is no specific law on the same in Monaco. However, any such claims would be dealt with under the Monegasque Civil Code on the guarantee against hidden defects likely to compromise the use of the thing sold[439]. Also, Monaco is in the process of drafting a legislative proposal for a special system of liability for defective products to be borne by producers/manufacturers if the product is the cause of personal injury or material damage.

In the Netherlands, during the summer of 2015, the road traffic rules were reviewed and amended to support AVs. The new amendment facilitated the testing of AVs on open roads after obtaining a permit from the competent authority[440]. Basis subsequent legislations the Dutch government on 01 July 2019 permitted testing of AVs on public roads under the Experimental Law on Self-Driving Vehicles, but under stringent conditions and subject to permission from various authorities like Police, road officials, the Netherlands Vehicle Authority (RDW), and the Dutch Institute for Road Safety SWOV[441].

---

436. Id.

437. Id.

438. Id.

439. Id.

440. Towards a European Regulation of Autonomous Vehicles - EU Perspectives and the German Model, Supra Note 442, at 120.

441. Self-driving vehicles Mobility, public transport and road safety | Government.nl, https://www. government.nl/topics/mobility-public-transport-and-road-safety/self-driving-vehicles#:~:text=The%20 Dutch%20cabinet%20has%20adopted,driving%20vehicles)%20removes%20legal%20impediments. (Last visited March 12, 2021).

The 2019 law permitted the testing of AVs without a driver sitting inside the vehicle, subject to the condition that the tested vehicles had a driver sitting outside the vehicle and were ready to control the AV if required. The 2019 law also permits moving motorway roadblocks with remote drivers

As per the 2019 law, the tests must demonstrate that the traffic safety risks are minimized, and the remote safety driver has a valid driving license. The test approval also carries with it a restricted time frame and the feedback on the tests must be given to the RDW[442]. Presently there are many cars on the road which have multiple functionalities like cruise control, automated parking, and automated lane-keeping systems. But these cars cannot be categorized as full AVs which have the capability to self-drive and thereby guarantee better traffic safety and reduction in the number of accidents. The Netherlands government has been quite clear since the beginning of the thought process on permitting testing of AVs that infrastructure including roads has to be made futureproof. They have been working on that in tandem with AV testing. The Netherlands government is also very sure that the outcomes of the road-testing experiments will be used as the basis to review the existing legislation based on their adequacy analysis.

On 01 January 2018, Norway permitted the testing of AVs on its roads and simultaneously started working on truck platooning. In May 2018 a company called Stavanger received permission to run a driverless minibus. As per the initial Norwegian regulations every AV was mandated to have an employee on board to apply emergency brakes if required. Other restrictions were a passenger limit of six and a max speed of 7.5 miles per hour (12 kilometres per hour). Norway started pilot testing of AV taxis in 2019. Norway has received good feedback on AVs and people are not scared to use them[443]. In Oslo, the percentage of electric vehicles sold exceeds 50% and most of them are now getting prepared for autonomous driving. EVs are more popular in Norway. This is mostly because of the large tax advantages, free charging stations, access to bus lanes, exemptions from parking fees and tolls, and low-cost residential electricity.

---

[442] Green light for Experimental Law for testing self-driving vehicles on public roads News item | Government.nl, https://www.government.nl/latest/news/2019/07/02/green-light-for-experimental-law-for-testing-self-driving-vehicles-on-public-roads (last visited March 12, 2021)

[443] 2019 Autonomous Vehicle Readiness Index Supra note 277, at 71. As per the report, Norway has by far the highest market penetration of EVs in the index, contributing to its second place in the technology and innovation pillar

There are different yardsticks in Poland when it comes to SAE Levels 3-5 AV testing on public roads. Level 5 testing is not permitted as the rules in Poland mandate the requirement of a safety driver to take control. Level 4 testing is permitted subject to compliance with laid down procedures and the availability of a safety driver at all times to take control if required. Level 3 again is permitted for testing almost on similar lines to Level 4. As of date consumers are not allowed to ply AVs on the roads in Poland[444]. Poland has no specific liability regime for AVs. Hence the existing general liability laws that apply to all motor vehicles would apply to AVs. Under normal circumstances, the registered owner would be responsible for the harm brought on by his vehicle/AV. Exemption from liability will kick in only during force majeure or in the case where the fault lies with the person who was harmed. The maker may be responsible for product liability under the general liability laws and also could be made liable under liability damages for hazardous products, which are moveable goods and carry strict liability regardless of fault[445]. AV testing organisations in Poland must have insurance for civil liability arising in connection with the testing of AVs. Though Poland has no specific law concerning warranty claims on AVs, a claim can lie on the manufacturer based on general liability[446].

The Road Traffic Code of Portugal requires all vehicles to have drivers with valid legal driving licenses. Level 3-5 AVs are not allowed on Portugal roads except in line with various European projects[447]. Resolution of the Portugal Council of Ministers no. 29/2020 has passed orders for the creation and regulation of "Technological Free Zones". These free zones are meant for testing Level 3-5 AVs. However, while testing is allowed, plying of Level 3-5 AVs by consumers is currently not allowed in Portugal. Since there is no specific law in Portugal which permits SAE Levels 3-5 AVs, liability as of date would be determined based on the Portuguese Civil Code's general provisions on civil liability. Any damage brought on by the car would be the responsibility of the owner/driver. The owner may still be liable, in cases

---

[444] Autonomous Vehicles Law and Regulations in Poland [online] available at https://cms.law/en/int/expert-guides/cms-expert-guide-to-autonomous-vehicles-avs/poland (last visited March 2, 2021)

[445] Id.

[446] Id.

[447] Autonomous Vehicles Law and Regulations in Portugal [online] available at https://cms.law/en/int/expert-guides/cms-expert-guide-to-autonomous-vehicles-avs/portugal (last visited March 2, 2021).

where the owner is not the driver if the vehicle is driven under his direction and in his interests[448].

Concerning product liability, the general liability rules of Portuguese law would be applicable, and for product damage, the manufacturer might be held accountable, including damages caused due to defects in the AV. The Portuguese Road traffic legal framework does not specifically talk of vehicle insurance for AVs. Hence the regular rules which mandate insurance for all automobiles will apply to AVs which are being tested on the road. Similar to insurance there is no specific warranty claim legal provision for AVs and hence warranty claims will be based on general rules on product liability and warranties[449].

There are no specific legislations in Romania for AVs. Hence it is a debatable issue if the existing testing laws would apply to AV testing. As per the existing laws of Romania, vehicle testing permissions are granted only to a vehicle manufacturer or an authorised tester. Also, in addition, the Romanian traffic regulations treat leaving the steering wheel while driving as an offence which could be a fine between Euro 60 to 90. Hence a person testing an AV without holding a steering would be committing an offence per Romanian laws[450]. Since drivers cannot leave the steering wheel while driving as per Romanian laws, consumers cannot drive AVs on public roads in Romania. However, Romania has authorised the sale of vehicles equipped with lidar sensors for autonomous parking, making them usable by customers.

Specific specifications for automated cars and fully automated vehicles are provided by the Type-Approval Standards for Motor Vehicles Regulation (EU) 2019/2144 of the European Parliament and of the Council of November 27, 2019 (EU Regulation). This EU legislation is expected to be applicable from 2022[451]. Since Romania has no liability rules enacted specifically for AVs any liability for damages caused by an AV would be subject to both the general civil liability laws for causing the damage as well as the current general traffic liability laws. As per the existing liability regime[452], civil and criminal liability for an owner and driver is as per the existing civil and criminal law

---

[448]. Id.

[449]. Id.

[450]. Autonomous Vehicles Law and Regulations in Romania [online] available at https://cms.law/en/int/expert-guides/cms-expert-guide-to-autonomous-vehicles-avs/romania (last visited March 2, 2021).

[451]. Id.

[452]. Id.

respectively. Exceptions granted would be for reasons of force majeure and contributory negligence. Exceptions granted could be for reasons connected with the responsibilities of the car manufacturer and AV system provider. If the automated systems are operated by the driver/operator as intended, the driver may also assert a defence against carelessness. According to current civil law regulations, the manufacturer is liable for product liability in instances where flaws cause death, physical harm, or damage to goods. In addition, the manufacturer can also be made criminally liable for product defects as per the existing criminal law of Romania. In line with what exists for the manufacturer, the AV system provider can also be held liable for civil and criminal negligence regarding damages caused by the operating system.

Since there are no specific regulations for AVs under Romanian laws all AVs would be governed by the general automobile insurance requirements including compulsory car insurance requirements for third-party damages. As per Romanian law, the victims can file a suit against the vehicle insurer and hence Romanian law requires all drivers to carry general auto insurance. Any such claim would impact the future value of insurance policies for the vehicle involved in the incident[453]. As regards general liability for warranty claims, in the absence of specific laws for AVs, the general civil rules of liability would apply to car manufacturers for product defects resulting in death, bodily harm or damage.

As a part of the focus of the Slovak Government in building and developing smart cities, the state is also focusing on AVs. Slovakia has a website https://www.smartcity.gov.sk/index.html for publishing operational programmes for the development of smart cities[454]. To support AVs, Slovakia is planning to develop an AV testing track on one of its highways, post-adoption of new legislation for AVs. As of date, there is no legislation in place for AVs in Slovakia. Hence AVs if any permitted for plying or testing before the passing of new law for AVs would be governed by existing motor vehicle laws for insurance, and liability[455]. The infrastructure Ministry of the Republic of Slovenia has prepared an amendment to the existing Slovenian Road Traffic Rules (Zakon o pravilih cestnega prometa - ZPrCP), which authorises

---

[453.] Id.

[454.] Autonomous Vehicles Law and Regulations in Slovakia [online] available at https://cms.law/en/int/expert-guides/cms-expert-guide-to-autonomous-vehicles-avs/slovakia (last visited March 2, 2021).

[455.] Id.

the testing of AVs on public roads as long as certain requirements are met. However, testing of AVs (SAE Levels 3-5) is not yet explicitly permitted under currently valid legislation[456]. Some of the stipulated conditions for testing AVs are marking of testing areas, marking of AVs being tested, monitoring of the safety driver of the AV regularly and also recording the driver's activities, insurance coverage of the AV being tested and adequate information to Police and Traffic department about the AV testing[457].

As per existing legislation in Slovenia consumers are permitted to use AVs with features only up to SAE Level 2 (e.g., Autonomous Parking). As per Slovenian Road Traffic Rules Act, the driver of a two-track vehicle must hold the steering at all times else there would be a fine of EUR 300. As regards vehicle with autonomous parking is concerned, the driver must be in the vehicle when it is being parked but is not obligated to hold the steering wheel. However, the driver must operate the clutch and brake apart from refueling the vehicle. Since there are no special rules concerning liability for damages caused by an AV, the general rules of liability would apply to the user and manufacturer. Under Slovenian laws, a moving vehicle is considered a dangerous object. Slovenian Obligations Code (Obligacijski zakonik – OZ) provides for strict liability of the holder of a dangerous object. The only exceptions for the user are if the vehicle was (i) removed unlawfully from the user and the user was not responsible for the removal, (ii) entrusted to another person by the user. In these cases, the person who unlawfully took the vehicle from the user or was entrusted with the use of the vehicle shall be liable instead of the user. Rules of strict liability apply to the manufacturer and the AV software developer for any defect in the vehicle or software which results in damages[458].

In Slovenia, insurance is mandatory for any vehicle which is plied on the roads. There are no additional insurance requirements for AVs. Every AV manufacturer is required to have signed a liability insurance contract for damages brought on by the use of an AV, according to the proposed amendment to the Road Traffic Rules Act (Zakon o pravilih cestnega prometa - ZPrCP), prepared by the Ministry of Infrastructure of the Republic of Slovenia. In addition to the above, the insurance must also cover the testing

---

[456.] Autonomous Vehicles Law and Regulations in Slovenia [online] available at https://cms.law/en/int/expert-guides/cms-expert-guide-to-autonomous-vehicles-avs/slovenia (last visited March 2, 2021).

[457.] Id.

[458.] Id.

period for the AV[459]. The AV driver must while driving the AV at all times have the insurance policy with him in the vehicle. As per Slovenian laws, warranty claims can be made against the seller and manufacturer of the vehicle. In case of any malfunction in the vehicle, the buyer can ask the seller or manufacturer to repair or replace the vehicle free of cost. Warranty starts from the day of delivery to a stipulated period post-delivery. If the vehicle had any manufacturing defect, then that's not covered by warranty and the manufacturer will be subject to strict liability under the applicable rules of the Obligations Code[460].

Another country in Europe that is working hard to support the entry of AVs on its roads is Spain. This country has started modifying its rules for AVs and is also modifying appropriate insurance laws to provide an overall framework for AV technology to operate in Spain. As early as November 2015 the Direccion General de Trafico (DGT) through an instruction approved the regulation for the testing of AVs. This instruction covers AVs upto Level 5 which means fully autonomous vehicles are permitted to be tested on Spain roads[461]. DGT Technology of Spain and a tech company Mobileye collaborated on January 18 to prepare the requisite infra ecosystem and regulatory policies for AVs in Spain, to reduce road accidents[462]. The combined research to ascertain the degree of safety improvement caused by the functionality and precision of Mobileye 8 Connect will be a part of the collaboration to encourage adoption. Through this collaboration, Barcelona has now been transformed into a full-fledged AV testing laboratory. It has around 5000 plus vehicles installed with Mobileye 8 Connect technology. In addition to the testing and infra ecosystem, Mobileye and DGT have also agreed to collaborate for redefining the regulatory roadmap and make it future-proof to take care of AVs[463].

A proposal was submitted to the Swedish legislative assembly in April 2016 to make the Swedish Transport Agency, the agency which can authorise permits for on-road testing of AVs in Sweden. Since 2017 the above-mentioned

---

[459]. Id.

[460]. Id.

[461]. Autonomous Vehicles Law and Regulations in Spain [online] available at https://cms.law/en/int/expert-guides/cms-expert-guide-to-autonomous-vehicles-avs/spain (last visited March 2, 2021).

[462]. Green Car Congress [online] available at https://www.greencarcongress.com/2018/01/20180106-mob.html (last visited March 21, 2021).

[463]. The state of autonomous legislation in europe | autovista group, https://autovistagroup.com/news-and-insights/state-autonomous-legislation-europe (last visited March 22, 2021).

agency was declared as the competent authority to grant permits and supervise AV trials[464]. The permits granted for testing in Sweden stipulate that in the case of fully Autonomous Vehicles, the legal responsibility for the operation of AVs and the resultant damages caused by such AVs shall be on the party which obtained the testing permit. However, in the case of a vehicle operating at lower automation levels and with a driver in control of the vehicle, the driver is responsible for assuming all civil and criminal liabilities. The Swedish government has also started a new project supported by Volvo. The project is called Drive me – self-driving cars for sustainable mobility. This project is basically for carrying out a large-scale test of AVs with the Traffic support of the citizens[465].

As a rule, driving AVs (SAE Levels 3-5) is not permitted on public roads in Switzerland. However, conditional testing can be permitted. As per Article 106 para 5 of the Federal Road Traffic Act (SVG), the Federal Council has the power to permit the testing of pilotless AVs. The testing if permitted has to be strictly carried out on the laid down parameters and routes[466]. Some of the conditions for grant of permission to test AVs are that the testing must be towards exploring and arriving at new findings concerning the AV technology, and Testers to explain the backup solution to make the process compliant, in case legal requirements are not met during initial testing. However, the testing of the vehicle is restricted to defined perimeters and requires the consent of the respective owners of the roads[467].

The permission granted for testing is time-bound and post-completion of the test the testing agency/organisation has to mandatorily submit a final report with the Federal Roads Office (FEDRO)[468]. The Federal Roads Office in Switzerland is working on traffic regulation and registration issues to allow all kinds of AVs to move on their roads. The legislation is in the process of being drafted and amongst the many changes, this also includes adaptions of

---

[464.] Automated Vehicles [online] available at https://www.transportstyrelsen.se/en/road/Vehicles/self-driving-vehicles/ (last visited March 22, 2021).

[465.] Self-Driving Cars for Sustaining Mobility [online] available at https://unece.org/fileadmin/DAM/trans/events/2014/Joint_Belgium-UNECE_ITS/03_ITS_Nov2014_Anders_Eugensson__Volvo.pdf (last visited March 21, 2021).

[466.] Autonomous Vehicles Law and Regulations in Switzerland [online] available at https://cms.law/en/int/expert-guides/cms-expert-guide-to-autonomous-vehicles-avs/switzerland (last visited March 2, 2021).

[467.] Id. Prior to testing there must be in place a Liability insurance for 100 million Swiss francs.

[468.] Id. More details on the testing and documentation are available at: https://www.astra.admin.ch/astra/de/home/themen/intelligente-mobilitaet/pilotversuche.html

the SVG or the Federal Act on Data Protection. As regards liability Swiss laws handle it in three buckets viz Owner, Operator and Manufacturer[469]. As per Article 58 para 1 SVG, the owner of a motor vehicle is liable for the damages caused by the operation of such vehicle, even though the vehicle was being driven by himself or someone else. This is basically "No Fault Liability". Akin to normal motor vehicles an AV owner is also required to have motor vehicle liability insurance ("*Motorfahrzeughaftpflichtversicherung*", Art. 63 SVG), which covers any damage caused while a motor vehicle is operated. As regards AVs there is no specific liability-related legislation as of the date and hence the general liability rules would apply for damages caused by an AV.

As regards vehicle operators, their liability is assessed according to Article 41 para 1 of the Swiss Code of Obligations (CO). Article 41 CO constitutes a fault-based liability. According to the current Article 31 para 1 SVG, which also applies to AVs (as there are no AV-specific laws for liability), at all times the operator must have control of the vehicle[470]. Hence for any failure by the AV operator to comply with Article 31 para 1 SVG, the AV operator will be held accountable for all damages the AV causes and will be held liable for their cost. As of date in Swiss parlance, only owners and operators' liability exist. According to Article 197 para 1 CO, there is a general warranty of the seller for the quality of sold goods. However, this warranty only applies to the seller and not to the manufacturer in general. In most cases, the manufacturer will not be the seller. But once AVs are on road then the product liability of the manufacturer under the Article 1 Product Liability Act would also be a part of the question of liability ("*Produktehaftpflichtgesetz*", PrHG;)[471]. Following the adoption of AVs, there may potentially be circumstances that call for joint culpability of the vehicle maker, owner, and operator under SVG Article 60 Paragraph 1. Under present Article 65 SVG, the grieving party can directly pursue the case with the insurance company.

Presently, Turkey has no regulations governing AVs. As per existing laws in Turkey, even vehicles being used for test driving must have a temporary registration and the number must be affixed to the vehicle. In 2019 Turkey announced AV testing by AVL Turkey, an international engineering company

---

[469.] Id.

[470.] Swiss Code of Obligations [online] available at https://max-eup2012.mpipriv.de/index.php/Swiss_Code_of_Obligations_(OR) (last visited March 21, 2021).

[471.] Id.

post-2020. The testing has not started yet[472]. As per existing motor vehicle laws in Turkey, every person who operates a car must have a driver's license and the same analogy would apply in the case of AVs too. Hence an AV driver also would need to have a driving license. The existing Turkish law governing automobiles defines a person who drives as a driver and owner as an operator[473]. According to Article 85 of the Law, both the operator and the driver are responsible for any accident that happens while using a motor vehicle. This is based on Strict Liability principles. In the absence of any separate law for AVs, the same strict liability principles would apply while handling liability cases of AVs. Since the existing law does not cover manufacturers, the liability of manufacturers would be assessed under the Turkish Code of Obligations and the Consumer Protection Law. For damages caused due to defects in the vehicle, the Consumer Protection Law may allow for the maker to be held accountable. and asked to provide a free repair of the vehicle or replace the vehicle.

As mentioned earlier, due to the absence of AV-specific laws, any criminal liability will be handled as per existing Turkish criminal law. Under the existing Turkish criminal law, only natural persons can be held criminally liable and not legal persons like manufacturers. Hence programmers of the AVs, drivers and users of the vehicles could be held criminally liable if they negligently or intentionally commit criminal offences (taksirli suç ve kasti suç)[474]. The court is likely to also look at the SAE level of the AV before deciding on the liability. For AVs at SAE levels 3-5, the liability and connection of manufacturers and programmers with the cause of damage will be assessed. However, for AVs at SAE levels 1-3, the possibility of driver liability will increase. Concerning vehicle insurance, the general Turkish laws would apply, and the owner must procure compulsory traffic insurance. As regards warranty claims, the general principles of the Turkish Code of Obligations would apply. In addition, manufacturers and importers of vehicles are obliged to issue guarantee certificates to consumers under the Consumer Protection Law. With the above principles in mind, it can be safely concluded that for a defective vehicle, a consumer is entitled to get the vehicle repaired by the manufacturer in cases

---

[472] Autonomous Vehicles law and regulation in Turkey cms expert guides, https://cms.law/en/int/expert-guides/cms-expert-guide-to-autonomous-vehicles-avs/turkey (last visited March 22, 2021)
[473] Id.
[474] Id.

where the manufacturer is obliged to repair the vehicle under a guarantee certificate. The seller, manufacturer, and importer may all be held jointly and severally liable if the manufacturer rejects the consumer's claims.

As of date, Ukraine has no separate or special laws for AVs, hence their usage on Ukraine roads is not explicitly permitted. The procedure for approving the construction of vehicles, their parts, and equipment as well as the procedure for maintaining the register of vehicle type certificates and equipment as well as certificates of conformity of vehicles or equipment issued by manufacturers, which was approved by the order No 521 of the Ministry of Infrastructure dated 17 August 2012, must be followed by the test vehicles used on the public road network to carry out the tests to introduce the vehicles into mass production[475]. Since there are no separate AV-related laws there is little ambiguity in using the "test vehicle" option to test AVs in Ukraine. As per general regulation for Ukrainian traffic rules (approved by the Regulation of the Cabinet of Ministers dated 10 October 2001 No. 1306), a human driver is mandatory to operate a vehicle[476]. The driver must be attentive at all times and never be distracted from driving. Because of the foregoing use of AVs (SAE levels 3 – 5) on public roads in Ukraine would be non-compliant with Ukrainian traffic rules.

As AVs are not permitted to be driven in Ukraine there are no past precedence of any decision by Ukrainian courts on liability for damages caused by AVs. Theoretically, there could be two ways to decide on the allocation of damages, one is that the legal owner of the vehicle could be made liable as per the Civil Code of Ukraine, or as per the Law of Ukraine for any liability on account of damages attributable to product defect, the manufacturer, or the importer on record of the AV could be made liable for the defected AV and all damages caused by this AV. However, the final method to be used in the case of AVs will have to be clarified by court decisions or specific laws. As regards insurance of AVs in the absence of any AV-specific laws, should there be a requirement then the courts will have to look at the general rules of Ukrainian insurance laws. Concerning product defects in the case of AVs as per the general rules of defect of products claims, the manufacturer can be held liable. There is no limitation on the manufacturer's liability under the

---

[475.] Autonomous Vehicles law and regulation in Ukraine cms expert guides, https://cms.law/en/int/expert-guides/cms-expert-guide-to-autonomous-vehicles-avs/ukraine (last visited April 2, 2021)
[476.] Id.

Commercial Code of Ukraine[477]. The customer will have the right to seek a proportional reduction of the price and/or seek free-of-charge elimination of defects within a reasonable time and/or seek reimbursement of costs for elimination of defects. If the manufacturing defects are of a serious nature defect, a customer may either terminate the sale purchase agreement and seek a refund or demand substitution of the vehicle.

The UK government has recently come up with a new department called the Centre for Connected and Autonomous Vehicles (CAV) and this department is working on bringing out appropriate legislation to permit the testing of AVs on UK roads. As regards AVs, since 2014, the British government has been funding more than 200 organisations on more than 70 projects. Most of these projects are aimed towards finding solutions to a variety of problems on connectivity and automation. These projects also cover many kinds of vehicles from small passenger-carrying pods to Heavy Goods Vehicles (HGV)[478]. The UK also has a government-supported centre of excellence called TSC and this centre has been a forerunner in the process of creating and delivering ambitious connected AV projects in the UK. TSC apart from supporting technological development for AVs is also supporting infrastructure development in the UK[479]. The UK government has also come out with a "Code of Practice for Automated Vehicle Trialling". This code provides details of government expectations from companies wanting to test AVs on UK roads. As per the above code, the responsibility of complying with all laid down legal requirements is that of the organisations carrying out AV trials[480]. There have been complications during compliance, especially concerning rules which were developed without insight into the AVs that would have to comply with those rules. Therefore, the Scottish Law Commission and the Law Commission of England and Wales are jointly evaluating the current legal framework for AVs on behalf of the UK Government.

[477.] Id.

[478.] Connected and Autonomous Vehicles: guidance for London trials [online] available at https://content.tfl.gov.uk/connected-and-autonomous-vehicles-guidance-for-london-trials.pdf (last visited March 21, 2021).

[479.] Market Forecast for connected and autonomous vehicles [online] available at https://assets.publishing.service.gov.uk/government/uploads/system/uploads/attachment_data/file/642813/15780_TSC_Market_Forecast_for_CAV_Report_FINAL.pdf (last visited March 21, 2021).

[480.] 91 Code of Practice: automated vehicle trialing [online] available at https://www.gov.uk/government/publications/trialling-automated-vehicle-technologies-in-public/code-of-practice-automated-vehicle-trialling (last visited March 21, 2021).

3.7 Automated Lane Keeping System (ALKS) Regulation

Under the aegis of the United Nations Economic Commission for Europe (UNECE), around 50 plus countries including Japan, South Korea and around 42 EU countries have agreed on a common regulation for AVs (ALKS) especially the kind of AV which can completely take over driving functions[481]. As per UNECE, this law is going to be the first international agreement that is binding on Level 3 AVs and above (driverless) thereby helping the world achieve safe and sustainable mobility for all. Japan, an Asian country had co-led the drafting along with Germany. Japan has assured to implement the regulations once it is in place. The European Commission (EC) was also another partner in the project apart from France, Canada and Netherlands. These restrictions would be applicable in the EU at a later time, according to the EC. Despite not being a member of the forum, the United States of America must abide by the rules if it wants to sell its AVs in Japan[482].

The ALKS[483] has a requirement to have a Blackbox in all AVs which can record every minute detailed recording of the AV's journey and how the AV was handled during the journey or an accident. The speed limit under ALKS has to be upto 60 kilometres per hour or 37 miles per hour for all AVs[484]. ALKS can be activated only when the safety driver is behind steering with a seatbelt on and only on roads which has a divider and where cycling and walking are forbidden. If using ALKS, screens monitoring other activities in full AV mode are to be switched off as soon as the driver takes control of the vehicle[485]. The ALKS regulation also mandates that AV manufacturers introduce Driver Availability Recognition Systems. This will help monitor the safety driver's capability to take back control of the vehicle. The control could be taken over by the driver in many ways which could include methods like spotting eye blinking and closure. ALKS must also abide by regulations governing software updates and cyber security[486].

[481.] Countries agree regulations for automated driving Tech Xplore - Technology and Engineering news, https://techxplore.com/news/2020-06-countries-automated.html (last visited April 2, 2021).

[482.] Id.

[483.] Compliance with new Automatic Lane Keeping System Regulation, ALKS, https://www.tuvsud.com/en-in/industries/mobility-and-automotive/automotive-and-oem/autonomous-driving/compliance-with-new-automated-lane-keeping-system-regulation (last visited April 12, 2021)

[484.] Id.

[485.] Id.

[486.] Id.

3.8 Important/landmark case laws on Autonomous Vehicles in APJC and EU

As of date, there have been six fatal car crashes involving Tesla cars with autopilot of which two have been in APJC - one in China and one in Japan.

3.8.1 Handan, China (January 20, 2016) Tesla Crash

Gao Yaning, a Tesla driver, was killed on January 20, 2016, in Handan, China, when his vehicle collided with a stationary truck. The Tesla car was plying, behind another car which suddenly changed to the right lane for avoiding a truck parked on the left shoulder which the Tesla car in autopilot mode did not detect. The deceased driver's family did file a lawsuit in September 2016 against Tesla to let the public know about the defects in Tesla's self-driving technology. Tesla responded saying the car was damaged heavily and hence telemetry could not be retrieved to understand if autopilot was engaged or not[487]. Six months later, Gao's father Gao Jubin sued Tesla, alleging that Tesla had oversold and neglected to disclose the limitations of its Autopilot system. He said that when his son passed away on the Beijing-Hong Kong-Macao Expressway in Hebei province, the automobile was in self-driving mode. At a hearing on September 20, 2016, Jubin asked the Beijing Chaoyang District People's Court to direct an impartial investigation into the crash. The lawsuit was stalled in 2018 due to issues in retrieving telemetry data. Later the deceased driver's family managed to retrieve the information and they sent that to Tesla, after which Tesla acknowledged that the accident occurred two minutes after the autopilot was turned on. The phrase Autopilot was later removed from Tesla's Chinese website.

3.8.2 Kanagawa, Japan Tesla Crash

On April 29, 2018, a Tesla car operating in Autopilot mode struck and killed a pedestrian in Kanagawa, Japan, who was 44 years old. The driver of the Tesla car was asleep at the time of the accident. A lawsuit has been filed in a federal

---

[487.] Tesla admits autopilot feature led to fatal China crash in 2016 [online] available at https://www.yicaiglobal.com/news/tesla-admits-autopilot-feature-led-to-fatal-china-crash-in-2016 (last visited March 24, 2021).

court in Northern California in April 2020[488]. According to the NTSB, Tesla's technology falls short of achieving full driver disengagement and is a poor indicator of engagement when it comes to tracking the driver's interactions with the steering wheel. The lawsuit[489] has apportioned the cause of the accident to a flaw in Tesla's autopilot technology, such as insufficient oversight of careless drivers. In a lawsuit filed in Tesla's home state of California, the victim's family claims that Tesla willfully disregarded the NTSB's recommendations for safety regulators[490].

3.8.3   *Arendal, Norway Tesla Fatal Crash*

On 29 May 2020, a Tesla car hit a truck driver who was walking alongside his semi-trailer on the road (which was partially off the road). The Tesla driver has been accused of causing a fatality negligently. One of the early witnesses in the trial who happened to be an expert witness has deposed that the vehicle was on autopilot when the accident occurred. A forensic scientist mentioned that the deceased truck driver was in the shadow of the trailer and hence less visible[491]. The driver was sentenced to three months' imprisonment in December 2020.

3.8.4   *Non-Fatal Tesla Crash (Taiwan, 01 June 2020)*

On 01 June 2020, a tesla went and hit an overturned cargo truck. While this was happening the traffic, cameras captured the entire accident. The motorist informed emergency personnel that the automobile was in Autopilot mode but that he or she was unharmed. According to reports, the driver admitted to police that he saw the vehicle and manually applied the brakes too late to prevent the collision[492]. As a driving assistance system, Tesla's Autopilot holds drivers accountable for accidents by requiring constant attention to the

---

[488.] Tesla Autopilot Technology Killed a Man in Japan, According to This Lawsuit [online] available at https://www.motorbiscuit.com/tesla-autopilot-technology-killed-a-man-in-japan-according-to-this-lawsuit/ (last visited July 7, 2021).

[489.] Id.

[490.] Id.

[491.] Norway Today [online] available at https://norwaytoday.info/news/tesla-driver-charged-with-negligent-homicide-of-truck-driver-he-hit-while-using-auto-steering-on-e18/ (last visited March 21, 2021)

[492.] Tesla In Taiwan Crashes Directly into Overturned Truck, Ignores Pedestrian, With Autopilot On [online] available at https://www.forbes.com/sites/bradtempleton/2020/06/02/tesla-in-taiwan-crashes-

road. This statement of the driver[493] is also corroborated by the recorded video where one could see a puff of white smoke coming from the tyres. Although their gaze is not recorded like it is in some vehicles, drivers must occasionally twist the steering wheel to demonstrate that they are maintaining their grip on it. Although he applied the brakes moments before the collision, the driver may not have been paying much attention because the truck is as obvious an obstruction as a roadside object can be. In this case, the driver is definitely to blame. Tesla frequently claims that its systems are extremely close to being able to perform true full-self driving, yet a system that accomplishes this is far from being prepared for it.

3.9   Conclusion

Various researchers have predicted that AVs in APJC will possibly cross 24 million in 2024[494]. This growth as of now is being propelled by stepping up R&D in the AV segment in APJC. As is being understood EVs and AVs are growing together, and their technologies are expected to complement each other. The growth of EVs is another great factor which is helping the expansion of AVs. There are many countries in APJC like India, Singapore and Malaysia that are developing infrastructure which can support the smooth implementation of EVs, and this is expected to help the growth of AVs. Awareness and adoption of Advanced Driver Assisted Systems or ADAS in AVs is helping boost the growth of the AV market in APJC and ANZ. With the increase in safety norms across all countries in the region, by incorporating ADAS into their vehicles, automakers can support self-driving technology and win over their customers.

Many automobile manufacturers are joining hands with technology companies to bring innovative and advanced technology to their products in the AV space. There are many such manufacturers and more than 250 AV companies who are also partnering for the development of such technologies to ensure that driverless cars are a reality. These businesses are categorised by the AV industry ecosystem into four main groups: automakers and auto-parts

---

directly-into-overturned-truck-ignores-pedestrian-with-autopilot-on/?sh=46729fd658e5 (last visited July 7, 2021).

[493.] Id.

[494.] Autonomous cars to grow to 24 million units in Asia-Pacific by 2024, Supra Note 340, at 92.

suppliers, technology firms, service providers, and tech start-ups. It was in 2017 that Tesla's Chief Executive Officer (CEO) Elon Musk promised to deliver a Level 5 vehicle within two years, subject to software validation and legal regime. Elon Musk had also said that the suggested vehicle would be able to travel autonomously between Los Angeles and New York City[495].

General traffic legislation is covered by the 1968 Vienna Convention (VC) on Road Traffic. 38 Council of Europe member states has ratified and signed this convention. All drivers must be able to handle their vehicles at all times, according to Article 8, Paragraph 5 and Article 13 of the VC. To accommodate AVs, a new paragraph was inserted into Article 8 in 2016. After this addition, AVs and Semi AVs will be considered compliant with the terms of this convention if the driver can take over control of the AV/Semi-AV system or if it complies with the terms of the 1958 UNECE Agreement and the 1998 Global Technical Regulations (GTR) Agreement[496].

In many parts of the world, more so in Europe, vehicles regularly cross borders. Hence there is a great amount of visionary thinking that the future legal regime for AVs should be developed with a global vision and context in mind. Presently there are a lot of international bodies collaborating in this space. Mostly all the Council of Europe member States have their legal regime for regulating AVs and semi-AVs. Countries like Germany and UK have enacted AV-specific laws, while others have been able to use existing motor vehicle, insurance and liability laws for AVs and Semi AVs. Despite this, there are several relevant international and regional instruments. Insurance, product liability, international vehicle standards, criminal and civil penalties for traffic infractions, roadworthiness standards and procedures, consumer information and marketing standards, driver's licences, accident investigation processes, data protection laws, and regulations regarding the use of the vehicles themselves are just a few of the topics covered by the laws that are currently in place regarding vehicle safety requirements (rules of the road)[497].

[495.] Business Insider, India [online] available at https://www.businessinsider.in/science/three-years-ago-elon-musk-said-a-tesla-car-would-be-able-to-drive-itself-across-the-country-by-the-end-of-2017-but-that-still-hasnt-happened/articleshow/68396520.cms (last visited July 27, 2021).

[496.] COMMITTEE ON LEGAL AFFAIRS AND HUMAN RIGHTS LEGAL ASPECTS OF AUTONOMOUS VEHICLES, https://assembly.coe.int/LifeRay/JUR/Pdf/DocsAndDecs/2020/AS-JUR-2020-20-EN.pdf (last visited April 12, 2021)

[497.] Id.

Legal Liability is another major issue to be addressed for the smooth use of AVs. For the present manually driven vehicles the insurance companies while calculating the premium to be paid, ascertain the age of the driver, gender, experience and roads the vehicle will be driven for risk assessment. The human element as cause of the accident is always considered prime in such calculations when deciding liability for any accident including those due to drunken driving and over-speeding. To implement AVs, however, the laws must be changed to assess culpability in situations where there is no driver present or when a driver is present but depending on automated controls. China policymakers are looking at shifting liability for AVs being driven in self-driving mode, from driver to manufacturer.

It can be summarized that the technology for semi-AV and full AV is well-developed, and the stage is set for commercial deployment. However, each country has budgetary, policy and regulatory issues which must be solved so that AV deployments can happen in full steam. The speed of AV deployment in a country would depend on the way the Government regulates AVs in that country. Governments must support road trials through well-defined regulatory priorities and proper control of AV policies, and this will help speed up AV deployment.

* * *

CHAPTER 4

CRITICAL ANALYSIS OF LEGAL REGIME FOR AUTONOMOUS VEHICLES IN THE REST OF THE WORLD

4.1 Background of Autonomous Vehicles

General Motors (GM) of the United States of America created an AV to demonstrate its vision of AVs for the world in an exhibition in 1939. This vision saw the light of the day twenty years later through a prototype model which included smart highway systems and autonomous vehicles[498]. To date, many manufacturers in the USA are working on AVs. As in the USA, manufacturers and AI technology innovators in Europe are working towards bringing out AVs for consumer use. There are some countries in Europe which are at an advanced stage of testing AVs while many are still in the preliminary stages only. APJC is also not behind in the development of AVs. There is a great amount of time and investment being spent by most nations towards Research and Development (R&D) in this fast-growing sector. There are reports that state that Asia Pacific Japan China (APJC) region will probably have 24 million AVs by 2024. Apart from AVs another big thing on which

---

498. History of the Autonomous Car, TitleMax (2020), https://www.titlemax.com/resources/history-of-the-autonomous-car/#:~:text=History of Autonomous Cars, made this concept a reality (last visited January 10, 2021).

nations in APJC are focusing is climate change and towards this Electric Vehicles or EVs are also an important aspect of R&D[499]. The current focus is more on AVs based on the EV platform.

To ensure the smooth adoption of AVs in our day-to-day life there is a need to have the right legal rules and regulations in place. These laws have to be written from a practical perspective and must cover all aspects of AVs be its driverless vehicles, Ethics for decision-making when choosing one of the two or more options in cases where collusion is imminent and Data Protection aspects. Upto November 2020, thirty-eight states in the USA along with the District of Columbia (DC) have passed laws/executive orders for testing and plying of Autonomous Vehicles (AVs)[500]. Other states either have them in the pipeline or are yet to decide on them. As in the USA, manufacturers and AI technology innovators in Europe are working towards bringing out AVs for consumer use. Even governments across many countries in Europe are supporting these efforts through research funding and appropriate legislation to permit trials. When one talks of Europe and automobiles, one country that rings a bell is Germany. As expected, Germany has also been the pioneer nation in Europe to come up with a law in 2017 which permitted AVs to be tested and operated on public roads[501]. However, this was subjected to certain conditions. One of them was the mandatory requirement to have a safety driver in the car[502] who could control the car and the steering wheel on a required basis. A motor vehicle with autonomous driving capabilities can carry out the driving task independently within a predetermined operating territory without a person operating the vehicle, according to Section 1e, Paragraph 2 of the German Autonomous Driving Act. The German regulation of 2017 apart from creating a roadmap for AVs also worked to solve liability issues[503].

Around May 2018, the European Commission published a report titled "On the road to automated mobility: An EU plan for the mobility of the future". This document listed out a roadmap to outline the ambition of the

---

[499]. Autonomous cars to grow to 24 million units in Asia-Pacific by 2024, Supra Note 340, at 92.

[500]. Autonomous Vehicles, GHSA, https://www.ghsa.org/taxonomy/term/521 (last visited January 10, 2021).

[501]. Germany: Road Traffic Act Amendment allows driverless vehicles on public roads [online] available at https://www.loc.gov/item/global-legal-monitor/2021-08-09/germany-road-traffic-act-amendment-allows-driverless-vehicles-on-public-roads/ (last visited March 2, 2021.

[502]. Id.

[503]. Tilburg University Autonomous Vehicles Regulation in Germany and the US and its impact on the German car industry [online] available at http://arno.uvt.nl/show.cgi?fid=149595 (last visited March 2, 2021).

EU towards becoming a world leader in deploying AVs. The 2017 German Ethics Committee came up with a report containing 20 different principles on ethical programming in AVs[504]. Some of these principles included aspects concerning using networked AVs as ethically necessary only if they cause fewer accidents than human drivers. Many automobile manufacturers in APJC and ANZ are collaborating with a broad spectrum of tech companies to promote advancement in their AV solutions. Apart from this, these auto companies are also mutually collaborating amongst themselves to control costs and bring about better efficiency. Reports are stating that Asia Pacific Japan China (APJC) region will probably have 24 million AVs by 2024. Apart from AVs, another big thing on which nations are focusing is climate change and towards this Electric Vehicles or EVs are also an important aspect of R&D[505].

4.2 Overview of Legal Regime for AVs in the Rest of the World (other than India)

The legal regime for AVs is at an advanced stage in the USA followed by the EU and then the APJC region. The USA even has detailed laws for automated truck platooning which is not seen elsewhere. There are some countries which permit AV testing but have no AV laws and hence fallback on existing conventional laws for handling any legal issues arising out of the use of AVs. There are some similarities in some countries and some gaps between them. All these are discussed in the subsequent paragraphs.

4.2.1 No AV Laws

There are many countries in the world which have no laws for AVs and some of these prominent ones are Slovakia, Indonesia, Malaysia, Philippines, Vietnam, Australia, and New Zealand. In Slovakia, if AVs as on date are permitted for plying or testing before the passing of new AV law, they would be governed by existing motor vehicle laws concerning insurance, liability[506]

[504.] The German Ethics Code for Automated and Connected Driving [online] available at https://www.researchgate.net/publication/320011270_The_German_Ethics_Code_for_Automated_and_Connected_Driving (last visited March 2, 2021).

[505.] Autonomous cars to grow to 24 million units in Asia-Pacific by 2024, Supra Note 340, at 92.

[506.] Autonomous Vehicles Law and Regulations in Slovakia [online] available at https://cms.law/en/int/expert-guides/cms-expert-guide-to-autonomous-vehicles-avs/slovakia (last visited March 2, 2021).

and driver's license. The same is the case with Malaysia[507], Vietnam, Australia, and Indonesia. The absence of AV laws in the Philippines is a worry for their insurance industry especially regarding the potential repercussions once AV technology is implemented. Due to this reason the government has received a warning from the Filipino Insurance Commission to carry out a comprehensive overhaul of certain laws to bring about a cohesive environment for AVs to ply in the Philippines[508]. The existing motor vehicle laws in New Zealand can only control automobiles equipped with level 2 automated features like lane-keep assist, automatic emergency braking, and collision mitigation systems. Beyond Level 2 the existing laws cannot work, and hence New Zealand is working on new laws[509].

The Bulgaria Road Traffic Act 1999 (Road Traffic Act)[510] does not permit AV testing on public roadways in Bulgaria. There are no specific legislations in Romania for AVs. As of date, Turkey has no regulations governing AVs. As per existing laws in Turkey, even vehicles being used for test driving must have a temporary registration and the number must be affixed to the vehicle. Similarly, Ukraine has no separate or special laws for AVs, hence their usage on Ukraine roads is not explicitly permitted. Since there are no separate AV-related laws there is little ambiguity in using the "test vehicle" option to test AVs in Ukraine.

Analysis of the above reveals that amongst the countries which have no AV laws Slovakia is the only one which specifically calls out AVs being governed by existing motor vehicle, product liability and insurance laws. New Zealand permits only AVs upto SAE Level 2 to ply on their roads. The Slovakia model is an example for countries which want to adopt AVs but has a long law-making process. That way they can ensure the adoption of modern technology.

---

[507] Abu Kassim et al., *Is Malaysia* Ready *to Adopt Autonomous Vehicles?*, 3 Journal of Society of Automotive Engineers Malaysia 84–88 (2019).

[508] BCDA to pilot autonomous vehicles in New Clark City for SEA Games [online] available at https://bcda.gov.ph/news/bcda-pilot-autonomous-vehicles-new-clark-city-sea-games (last visited March 31, 2021).

[509] Current Regulatory Settings for Autonomous Vehicles [online] available at https://www.transport.govt.nz/about-us/what-we-do/queries/current-regulatory-settings-for-autonomous-vehicles/ (last visited March 27, 2022).

[510] Road Traffic Act [online] available at http://www.lawyersbulgaria.org/content/2011/603/road-traffic-act/ (last visited March 21, 2021).

4.2.2 *Vehicle Platooning*

Truck platooning is the merging of two or more vehicles in a convoy using networking technology and autonomous driving support systems. When these vehicles are connected for some part of a journey, say on a highway, they automatically maintain a predetermined, tight distance between each other. Kentucky, Mississippi, Florida, and Nevada have motor vehicle rules which permit vehicle platooning. As per Section 2 (39) of the Kentucky bill SB 116,[511] a platoon is defined as a pair of commercial motor vehicles moving together at electronically coordinated speeds and following distances that are closer than those typically permitted by subsection (8)(b) of Section 3 of the bill. As per Section 63-3-103, Mississippi Code of 2019[512], several different motor vehicles travelling in tandem at electronically controlled speeds with closer following distances are referred to as a platoon. As per Section 2 of Nevada Assembly Bill 69[513], driver-assistive platooning technology is that which enables two or more trucks or other motor vehicles to travel on a roadway at electronically coordinated speeds in a coordinated manner at a following distance which is closer than would be fair and prudent without the use of the technology.

There are a few states in the USA that mandate proper permission before plying vehicles in platoon format. In Arkansas, a person requiring plying such truck platooning systems must seek permission for platoon operations from the State Highway Commission[514]. Kentucky through SB 116 (March 2018) has permitted platoon plying of commercial vehicles after providing adequate notification to the Kentucky State Police and the Department of Vehicle Regulation (DVR). One of the mandatory ask in this bill is the requirement to have a driver with a valid commercial driver's license in each vehicle which is part of a platoon[515]. Louisiana vide HB 308 of May 2018 has permitted

---

[511.] Kentucky SB 116 [online] available at https://trackbill.com/bill/kentucky-senate-bill-116-an-act-relating-to-the-operation-of-a-commercial-motor-vehicle/1541369/ (last visited March 31, 2021).

[512.] 2019 Mississippi Code [online] available at https://law.justia.com/codes/mississippi/2019/title-63/chapter-3/article-3/vehicles-equipment-and-the-like-defined/section-63-3-103/ (last visited December 22, 2020).

[513.] Government Technology [online] available at https://www.govtech.com/fs/more-states-explore-truck-platooning-technology-and-regulations.html (last visited December 22, 2020).

[514.] What is truck Platooning (2017), https://www.acea.be/uploads/publications/Platooning_roadmap.pdf (last visited December 22, 2020).

[515.] Kentucky SB 116 [online] available at https://apps.legislature.ky.gov/law/acts/18RS/documents/0033.pdf (last visited February 23, 2022).

commercial vehicles to operate in a platoon[516] following endorsement by the Departments of Public Safety and Corrections and Transportation and Development[517]. Tennessee permitted the operation of vehicles in a platoon in April 2017, with requisite automated safeguards following the operator's notification to the Department of Transportation and Safety[518].

Some of the states in the USA also have exempted their existing "Too Close" regulatory restriction for vehicles plying in a platoon. California in 2017 approved drivers to maintain a safe distance of no more than 100 feet between any vehicle or group of vehicles when operating motor vehicles.[519]. The too-close provisions of 300 feet in case of vehicle platooning was exempted by Indiana bill no HB 1341[520]. Vide Mississippi Bill no 1343 passed in April 2018 non lead vehicle in a platoon was exempted from the too closely following law[521]. On 21 July 2017, vide Bill no 469[522] The following too-close law was changed to permit vehicle platooning after North Carolina established a statute regulating the operation of fully autonomous cars[523]. Vide Bill no 4059 passed in March 2018, Oregon[524] exempted the driver of a vehicle who was part of a connected automated braking system in a platoon from the too-close law. On 01 July 2017 Georgia passed a bill exempting the prohibition of following too closely by non-leading vehicles in a convoy, travelling in the same lane and employing technology to communicate amongst vehicles to coordinate their motions[525].

---

[516] Louisiana House Bill No 308 [online] available at https://custom.statenet.com/public/resources.cgi?id=ID%3Abill%3ALA2018000H308&ciq=ncsl&client_md=569a3280675b8f459612254d497b0650&mode=current_text (last visited December 26, 2020).

[517] Id.

[518] Vehicle Platooning [online] available at https://www.tn.gov/tdot/transportation-freight-and-logistics-home/vehicle-platooning.html (last visited December 28, 2020). Platooning is the linking of two or more vehicles in a convoy using wireless communications and sensor technology.

[519] BILL TEXT - AB-669 Department of Transportation: Motor Vehicle Technology testing., https://leginfo.legislature.ca.gov/faces/billNavClient.xhtml?bill_id=201720180AB669 (last visited December 22, 2020).

[520] Colin Wood, Slow and steady, Indiana readies autonomous vehicle framework StateScoop (2018), https://statescoop.com/slow-and-steady-indiana-readies-autonomous-vehicle-framework/ (last visited December 24, 2020).

[521] HB No 1343 [online] available at http://billstatus.ls.state.ms.us/documents/2018/html/HB/1300-1399/HB1343SG.htm (last visited December 26, 2020).

[522] General Assembly of North Carolina - House Bill No 469, https://legiscan.com/NC/text/H469/2017 (last visited December 28, 2020).

[523] Id.

[524] Legislative Assembly OF Oregon House Bill No 4059, https://legiscan.com/OR/text/HB4059/2018 (last visited December 28, 2020).

[525] State of Georgia House Bill No 472, https://www.legis.ga.gov/api/legislation/document/20172018/170675 (last visited December 24, 2020).

Wisconsin through Act 294 of 2017, provided an exception for platoons from the 500 feet minimum distance law for trucks weighing more than ten thousand pounds[526]. South Carolina on 19 May 2017 through bill no H3289 declared that the law regarding minimum following distance[527] does not apply to vehicles travelling in a platoon. This was in sync with a similar law in place in many states in the USA. As per the bill having appropriate consideration for the speed of such vehicles, the volume of traffic on the road, and the state of the roadway, the driver of a motor vehicle shall not follow another vehicle more closely than is reasonable and sensible.

Texas[528] permitted the usage of an electrically connected and synchronised braking system for enabling vehicles to maintain the appropriate distance between them[529] through HB 1791 on 01 September 2017. Section 545.062, Transportation Code of Texas was amended by adding Subsection (d) to read as follows:

"(d)An operator of a vehicle equipped with a connected braking system that is following another vehicle equipped with that system may be assisted by the system to maintain an assured clear distance or sufficient space as required by this section"[530] (Amended vide Texas HB 1791).

Apart from the above states in the USA, Norway is the only non-US country that has permitted testing of AVs on its roads and simultaneously started working on truck platooning[531] on 01 January 2018.

Analysis of all the above legal provisions on truck platooning in different jurisdictions indicates that the first and foremost requirement for truck platooning is to check if it is permissible in the jurisdiction where the trucks are to be plied and if yes, then appropriate permission is to be sought. The second important aspect is different jurisdictions have different clear distance rules between trucks and that needs to be duly complied with. The kind of driving license requirement for truck platooning is specified for each jurisdiction, which also needs to be complied with.

[526.] Report of the Governor's Steering Committee on Autonomous and Connected Vehicle Testing and Deployment [online] available at https://wisconsindot.gov/Documents/about-wisdot/who-we-are/comm-couns/av-final-report-062918.pdf (last visited December 30, 2020).

[527.] 2017-2018 BILL 3289: Safe following distance, https://www.scstatehouse.gov/sess122_2017-2018/bills/3289.htm (last visited December 28, 2020).

[528.] State of Texas HB 1791, https://capitol.texas.gov/tlodocs/85R/billtext/pdf/HB01791F.pdf#navpanes=0 (last visited December 28, 2020).

[529.] Id.

[530.] Id.

[531.] 2019 Autonomous Vehicle Readiness Index Supra note 277, at 71.

4.2.3 *Policies and Executive Orders on various aspects of AVs*

Most of the states in the USA require permission to be obtained before testing AVs. Some have also prescribed/approved testing agencies and other conditions. California has permitted AV testing only by Contra Costa Transportation Authority[532] and Livermore Amador Valley Transit Authority[533]. However, testing is permitted only at specified places with a vehicle speed of 35 miles or below after submitting a surety bond of $5 million[534]. Colorado laws require ADS to control motor vehicles while DC laws mandate the Department of Motor Vehicles to establish safe operating protocols for AVs to ply in the District of Columbia[535]. Florida had as early as July 2012, declared its support for AV testing and operations[536], with a mandatory, ask for AVs to maintain federal safety standards. Maine in January 2018 established a Highly Automated Vehicles (HAV) Advisory Committee[537] and permitted testing of AVs only with permits. To support this Maine in April 2018 established a Commission on AVs for allowing the testing, demonstration, and deployment of Automated Vehicles[538]. In December 2013 vide Senate Bill 663[539] and in March 2014 vide Senate bill 169, Michigan permitted testing of automated vehicles under laid down conditions[540]. Section 4 of Nevada SB 313[541] requires AVs to meet all applicable federal standards and regulations before being permitted to register in the state.

---

[532] Assembly Bill No 1592 [online] available at http://www.leginfo.ca.gov/pub/15-16/bill/asm/ab_1551-1600/ab_1592_bill_20160830_enrolled.pdf (last visited February 23, 2022).

[533] BILL TEXT - AB-1444 Livermore Amador Valley Transit Authority: demonstration project., https://leginfo.legislature.ca.gov/faces/billTextClient.xhtml?bill_id=201720180AB1444 (last visited December 22, 2020).

[534] Assembly Bill No 1592 [online] available at http://www.leginfo.ca.gov/pub/15-16/bill/asm/ab_1551-1600/ab_1592_bill_20160830_enrolled.pdf (last visited February 23, 2022).

[535] DC Act 19-643 [online] available at https://lims.dccouncil.us/downloads/LIMS/26687/Signed_Act/B19-0931-SignedAct.pdf (last visited February 23, 2022).

[536] An Attempt to Control What Controls Itself: Unraveling Florida's Autonomous Vehicle Laws [online] available at https://www.phelps.com/a/web/6rLYXJMb3axzKRFn2GR53m/ta-vol-38-n3-2019-dahdah-av-laws.pdf (last visited February 23, 2022).

[537] Maine Executive Order 2018-001 [online] available at https://www.maine.gov/mdot/autonomous-vehicles/docs/2018-001_HAVExecOrder.pdf (last visited February 23, 2022).

[538] State of Maine HP1204-LD1724 [online] available at https://mainelegislature.org/legis/bills/getPDF.asp?paper=HP1204&item=3&snum=128 (last visited December 26, 2020).

[539] State of Michigan Enrolled Senate Bill No 663, https://www.legislature.mi.gov/documents/2013-2014/publicact/pdf/2013-PA-0251.pdf (last visited December 26, 2020).

[540] State of Michigan Enrolled Senate Bill No 169, https://www.legislature.mi.gov/documents/2013-2014/publicact/pdf/2013-PA-0231.pdf (last visited December 26, 2020).

[541] State of Nevada Senator Bill No 313, https://www.leg.state.nv.us/Session/77th2013/Bills/SB/SB313_EN.pdf (last visited December 26, 2020).

On 23 January 2017 New York passed a bill[542] empowering the motor vehicle commissioner to approve AV testing. Under its first Automated-Connected-Mobile Action Plan, Austria has come up with a legal regime for testing AVs and doing research on AVs with an investment of around 25 million euros[543] and with over 300 AV professionals. Post an amendment in 2017 road testing of AVs was legalised in Hungary[544] subject to specific prior permission from the Ministry of Traffic. The Ministry can refuse permission if the proposed testing does not comply with certain legal and technical requirements. As of date consumers are not permitted to drive AV on the roads in Hungary. In February 2018, Italy passed its first law permitting the testing of AVs[545]. The law allowed testing only on specific roads. Further testing has to be authorised by the Appropriate Authority which is the Ministry of Infrastructure and Transport. As of right now, consumers in Italy are not allowed to utilise AVs (SAE Levels 3-5) on public roads.

As per Sovereign Order n° 7.489 of 27 May 2019 concerning plying of AVs on public roads, AVs are allowed to ply on the roads in Monaco for testing purposes with either full or partial delegation of driving on a public road. However, before conducting such testing permission from the Minister of State is to be obtained by the manufacturer according to the guidelines outlined in Ministerial Order No. 2019-512[546]. As of date, Monaco does not have any legal regime which permits the use of AVs by consumers on public roads.

The Netherlands has facilitated the testing of AVs on open roads after obtaining a permit from the competent authority[547]. The test approval also carries with it a restricted time frame and the feedback on the tests must be

---

[542] State of New York S 2005-C, A 2005-C, https://nyassembly.gov/2017budget/budget_bills/A3005C.pdf (last visited December 26, 2020).

[543] 2019 Autonomous Vehicle Readiness Index Supra note 277, at 71.

[544] Towards a European Regulation of Autonomous Vehicles - EU Perspectives and the German Model, Supra Note 442, at 120.

[545] Autonomous Vehicles Law and Regulations in Italy [online] available at https://cms.law/en/int/expert-guides/cms-expert-guide-to-autonomous-vehicles-avs/italy (last visited March 2, 2021)

[546] Autonomous Vehicles Law and Regulations in Monaco [online] available at https://cms.law/en/int/expert-guides/cms-expert-guide-to-autonomous-vehicles-avs/monaco (last visited March 2, 2021). As per 2019 order, the application for permission must mandatorily specify the level of automation.

[547] Towards a European Regulation of Autonomous Vehicles - EU Perspectives and the German Model, Supra Note 442, at 120.

given to the RDW[548]. Norway started pilot testing of AV taxis in 2019[549]. The Russian government, on 26 November 2018 issued regulations which permitted the testing of driverless vehicles on Russian roads. This was done vide Regulation no 1415[550]. The Russian regulations also mandate that every driverless vehicle must have a large black letter A mentioned on it to identify it as an AV. Direccion General de Trafico (DGT) of Spain in November 2015 through an instruction approved the regulation for testing of AVs. This instruction covers AVs upto Level 5 which means fully autonomous vehicles are permitted to be tested on Spain roads[551]. Swedish legislative assembly in April 2016 approved the Swedish Transport Agency as the agency to authorise permits for on-road testing of AVs in Sweden[552]. As per Article 106 para 5 of the Switzerland Federal Road Traffic Act (SVG), the Federal Council has the power to permit the testing of pilotless AVs. The testing if permitted has to be strictly carried out on the laid down parameters and routes[553].

The UK government has recently come up with a new department called the Centre for Connected and Autonomous Vehicles (CAV). To enable testing of AVs on UK roads, this department is attempting to develop the necessary legislation.[554]. The UK government has also come out with a Code of Practice for Automated Vehicle Trialling. This code provides details of government expectations from companies wanting to test AVs on UK roads. AV tests were first officially allowed by the Chinese Government only in 2018 as by then China had a large number of local AV companies[555] who were working hard to compete with global OEMs. The initial AV tests in China

---

[548.] Green light for Experimental Law for testing self-driving vehicles on public roads News item | Government.nl, https://www.government.nl/latest/news/2019/07/02/green-light-for-experimental-law-for-testing-self-driving-vehicles-on-public-roads (last visited March 12, 2021)

[549.] 2019 Autonomous Vehicle Readiness Index Supra note 277, at 71.

[550.] Government.ru. 2021. *О проведении эксперимента по эксплуатации на автодорогах высокоавтоматизированных транспортных средств.* [online] Available at: <http://government.ru/docs/34831/> [last visited June 28, 2021].

[551.] Autonomous Vehicles Law and Regulations in Spain [online] available at https://cms.law/en/int/expert-guides/cms-expert-guide-to-autonomous-vehicles-avs/spain (last visited March 2, 2021).

[552.] Automated Vehicles [online] available at https://www.transportstyrelsen.se/en/road/Vehicles/self-driving-vehicles/ (last visited March 22, 2021).

[553.] Autonomous Vehicles Law and Regulations in Switzerland [online] available at https://cms.law/en/int/expert-guides/cms-expert-guide-to-autonomous-vehicles-avs/switzerland (last visited March 2, 2021). g

[554.] Connected and Autonomous Vehicles: guidance for London trials [online] available at https://content.tfl.gov.uk/connected-and-autonomous-vehicles-guidance-for-london-trials.pdf (last visited March 21, 2021).

[555.] To Find China's Best Driverless Technology, Look in Silicon Valley [online] available at https://www.bloombergquint.com/business/to-find-china-s-best-driverless-technology-look-in-silicon-valley (last visited April 25, 2021"

were conducted by two Chinese start-ups Jingchi and Pony.ai. These tests were carried out in Guangzhou in 2018[556]. In PRC the person who applies for the testing has to have a registered legal status in PRC (Entity) concerning ICV for manufacturing, testing and R&D. According to the Road Traffic (Autonomous Motor Vehicles) Rules 2017 (2017 Rules), only those who have the permission of Singapore's Land Transport Authority (LTA) are allowed to test AVs in Singapore. The LTA is empowered to grant permission only in a specific area for the trial along with other conditions[557].

The Department of Highways and the Department of Land Transportation in Thailand are working with other government agencies to formulate plans for Thai roads to facilitate the smooth adoption of AVs in Thailand. 2017 saw the introduction of the first AV (car) in Vietnam[558]. Australian government started the process of phased transformation of existing driving laws in 2016 to enable the complete use of AVs[559]. An area of Christchurch in New Zealand post the 2011 earthquake was categorized for redevelopment and has been identified and approved by the New Zealand government as a location for AV trial. This place would also be known as Mobility Lab[560].

Analysis of the policies on AVs in different jurisdictions reveals that for successful AV implementation a country would have to adopt the following i.e., require AV testers to take due permission, prescribe specific areas for AV testing and if feasible only allow certified agencies to carry out AV testing. The Russian requirement for every driverless vehicle to be marked with a huge black letter A to distinguish it as an AV is a great way to inform pedestrians and other vehicles know of the AV on the road. The testing areas, speed limits etc are also specified in the government orders and approval for testing.

---

556. Id.

557. Road traffic Autonomous Motor Vehicles Rules 2017 [online] available at https://sso.agc.gov.sg/SL/RTA1961-S464-2017?DocDate=20170823 (last visited March 25, 2021).

558. Vietnam's first autonomous vehicle debuts, vietnamnews.vn, https://vietnamnews.vn/economy/914897/viet-nams-first-autonomous-vehicle-debuts.html (last visited May 10, 2021).

559. Office of Future Transport technologies launched [online] available at https://www.spatialsource.com.au/federal-office-of-future-transport-technologies-announced/ (last visited March 24, 2021).

560. Inside New Zealand Police's Mobility Innovation Lab and Experience Centre [online] available at https://blog.hexagonsafetyinfrastructure.com/inside-new-zealand-polices-mobility-innovation-lab-and-experience-centre/ (last visited June 23, 2021).

4.2.4 *Various Definitions of AVs*

Connecticut passed an act concerning AVs[561] in June 2017 and as per the act, a fully autonomous vehicle is a motor vehicle equipped with level four or five ADS. As per Section 1 (a) of the said act:

> "Fully autonomous vehicle means a motor vehicle that is equipped with an automated driving system, designed to function without an operator and classified as level four or level five by SAE J3016" (Connecticut SB 00260).

As per Section 1 of Nebraska Bill, no LB 989 of 2018:

> "Automated driving system means the hardware and software that are collectively capable of performing the entire dynamic driving task on a sustained basis regardless of whether it is limited to a specific operational design domain if any" [562] (Nebraska LB 989 of 2019).

The 2017 Georgia bill[563] describes a fully autonomous vehicle as one that has an ADS and is capable of carrying out all parts of dynamic driving without the assistance of a human driver within a constrained or unconstrained operational design domain. The definition further adds that when the ADS is working within its operational design domain, AV will never ask a driver to take up any part of the dynamic driving task. As per Section 4 (2) of Nevada SB 313[564], an Autonomous vehicle before being permitted to be tested must have means to engage and disengage the autonomous technology. There must be a visual indicator inside the vehicle to show that the AV technology is in operation, technology to alert the safety driver to command the autonomous vehicle manually in case a failure happens or is detected as likely to happen

On 23 January 2017 New York passed a bill[565] wherein, under Section 1(b), the phrase autonomous vehicle technology means the combined power

---

[561] An Act concerning Autonomous Vehicles, https://www.cga.ct.gov/2017/ACT/pa/2017PA-00069-R00SB-00260-PA.htm (last visited December 24, 2020).

[562] Nebraska Legislative Bill 989 [online] available at https://legiscan.com/NE/text/LB989/2017 (last visited December 26, 2020).

[563] State of Georgia Senate Bill No 219, https://www.legis.ga.gov/Legislation/20172018/170801.pdf (last visited December 24, 2020).

[564] State of Nevada Senator Bill No 313, https://www.leg.state.nv.us/Session/77th2013/Bills/SB/SB313_EN.pdf (last visited December 26, 2020).

[565] State of New York S 2005-C, A 2005-C, https://nyassembly.gov/2017budget/budget_bills/A3005C.pdf (last visited December 26, 2020).

of the hardware and software to consistently complete all or a portion of the dynamic driving task, where dynamic driving job refers to all real-time operational and tactical activities necessary to drive a car in on-road traffic, excluding strategic tasks like planning trips and choosing locations and waypoints.

The most prominent difference between the US regulations and that of Germany is in the definition of AVs. In the USA Level 5 AVs can be driverless. But as per StVG in Germany, there is a need to have a safety driver in all AVs and they must have the capability to take over control on a required basis[566].

Analysis of the definition of AVs across the world reveals that most of the definitions are almost similar and they talk about a vehicle equipped with an automated driving system that is intended to operate without a driver. The SAE levels are also often referenced in the definition, and a vehicle must be of SAE Level Four or Five to qualify as a fully autonomous vehicle. The Russian regulations also mandate that a big black letter A must be present on any autonomous vehicle to identify it as an AV. The most prominent difference between the US regulations and that of Germany is in the definition of AVs. In the USA Level 5 AVs can be driverless. But as per StVG in Germany, there is a need to have a safety driver in all AVs and they must have the capability to take over control on a required basis. Hence any country adopting AVs can use the US model or German model depending on how it fits into their country.

4.2.5   Insurance for AVs

There are various kinds of Insurance norms across the world when it comes to AVs. When we look at the laws in the USA, Connecticut Senate Bill No. 260[567] requires the vehicle operator to be seated in the driver's seat and to be insured for at least $5 million. Similarly in Florida[568], the driver of an AV would require adequate insurance. The quantum has not been specified.

---

[566] Germany: Road Traffic Act Amendment allows driverless vehicles on public roads [online] available at https://www.loc.gov/item/global-legal-monitor/2021 08 09/germany road traffic-act-amendment-allows-driverless-vehicles-on-public-roads/ (last visited March 2, 2021.

[567] An Act Concerning Autonomous Vehicles., https://www.cga.ct.gov/2017/ACT/pa/2017PA-00069-R00SB-00260-PA.htm (last visited December 24, 2020).

[568] An Attempt to Control What Controls Itself: Unraveling Florida's Autonomous Vehicle Laws [online] available at https://www.phelps.com/a/web/6rLYXJMb3axzKRFn2GR53m/ta-vol-38-n3-2019-dahdah-av-laws.pdf (last visited February 23, 2022).

Nebraska in April 2019 vide Bill no LB 989, allowed a driverless vehicle to operate on the roads of Nebraska[569] and mandated vide this bill[570] to operators of these AVs to produce proof of proper insurance. In 2019 one more bill was submitted in Nebraska for approval and that was LB 142. This bill mandated that manufacturers of automated driving systems are required to carry $1 million in liability insurance and $5 million in property insurance[571]. This bill was strongly opposed and got suspended due to the very high insurance amount. Alabama Joint Legislative Committee [572] in April 2016 mandated that AVs must be covered by insurance. However, akin to Florida, here also no amount was specified. As per UK motor vehicle laws, all AV testing[573] must have adequate Insurance coverage. The amended Bulgarian Insurance Code 2016 (Insurance Code) also mandates compulsory third-party liability insurance to cover the liability of the driver, the passenger, and the vehicle's owner, as well as of any person who drives or uses the motor vehicle on valid legal grounds[574].

In Italy, there is a mandatory requirement to have the AV insured for civil responsibility involving the AV, with a maximum coverage equivalent to four times the legal minimum for the identical kind of vehicle without the autonomous driving feature. AV testing organisations in Poland must have insurance for civil liability arising in connection with the testing of AVs. As per Russian regulations, all AVs must be insured and protection must be provided for at least 10 million rubles (approximately $150,000). There is no explicit obligatory AV insurance requirement in Hungary and hence the general mandatory requirement that applies to any vehicle will apply to AVs too. As regards insurance for AVs there is no specific law in Monaco. However as per Ministerial Order No. 2019-512 AVs being tested need specific Insurance. As per the existing Ordinance, No. 2.617 of 23/08/1961 compulsory insurance is

---

[569] Nebraska Legislative Bill 989 [online] available at https://legiscan.com/NE/text/LB989/2017 (last visited December 26, 2020).

[570] Id

[571] Nebraska senator floats bill to address driverless vehicle liability, Roads & Bridges, https://www.roadsbridges.com/nebraska-senator-floats-bill-address-driverless-vehicle-liability (last visited February 16, 2021).

[572] William Thornton | wthornton@al.com, Alabama Legislature preparing for self-driving cars AL (2019), https://www.al.com/news/anniston-gadsden/2019/03/how-the-alabama-legislature-is-preparing-for-self-driving-cars.html (last visited December 17, 2020).

[573] 2019 Autonomous Vehicle Readiness Index Supra note 277, at 71.

[574] Insurance Code [online] available at https://www.abz.bg/public/uploads/files/code_en.pdf (last visited March 21, 2021)

required for all motor vehicles[575]. Going by this law all AVs irrespective of their SAE level would be subject to the above-mentioned compulsory insurance obligation. The Portuguese Road traffic legal framework does not specifically talk of vehicle insurance for AVs. Hence the regular rules which mandate insurance for all automobiles will apply to AVs which are being tested on the road. Since there are no specific regulations for AVs under Romanian laws all AVs would be governed by the general automobile insurance requirements including compulsory requirements for third-party liability insurance.

In Slovenia, insurance is mandatory for any vehicle which is plied on the roads. There are no additional insurance requirements for AVs. Every AV manufacturer is required to have signed a liability insurance contract for damages brought on by the use of an AV, according to the proposed amendment to the Road Traffic Rules Act (Zakon o pravilih cestnega prometa - ZPrCP), prepared by the Ministry of Infrastructure of the Republic of Slovenia. In addition to the above, the insurance must also cover the testing period for the AV[576]. As regards insurance of AVs in the absence of any AV-specific laws, should there be a requirement then the courts will have to look at the general rules of Ukrainian insurance laws. In Vietnam all individual vehicles are required to be registered with the competent Police Department and Vehicle owners must ensure that the vehicles are appropriately insured[577].

The ICV being tested in PRC must possess requisite technical requirements and must mandatorily have a compulsory traffic accident liability insurance coverage of over Renminbi (RMB) 5 million per tested vehicle. As of date, there are no ICV-specific insurance laws in PRC and hence in case of any accident, the driver will be liable as per existing laws. Akin to other countries PRC also has a third-party (TP) liability insurance system. This TP insurance must be as per the law procured by the vehicle owner or user. Noncompliance

---

[575] Autonomous Vehicles Law and Regulations in Monaco [online] available at https://cms.law/en/int/expert-guides/cms-expert-guide-to-autonomous-vehicles-avs/monaco (last visited March 2, 2021). As per 2019 order, the application for permission must mandatorily specify the level of automation.

[576] Autonomous Vehicles Law and Regulations in Slovenia [online] available at https://cms.law/en/int/expert-guides/cms-expert-guide-to-autonomous-vehicles-avs/slovenia (last visited March 2, 2021).

[577] Decree No 116/2017/ND-CP [online] available at https://english.luatvietnam.vn/decree-no-116-2017-nd-cp-dated-october-17-2017-of-the-government-on-requirements-for-manufacturing-assembly-and-import-of-automobiles-and-trade-in-117725-Doc1.html#:~:text=Decree%20No.,automobile%20warranty%20and%20maintenance%20services (last visited March 24, 2021).

in this regard can result in the seizure of the vehicle and a fine equivalent to twice the premium for the minimum liability limit[578].

Most jurisdictions across the world where AVs are permitted to ply or tested require the AV to be insured for at least One Million USD to Five Million USD. Nebraska mandates that autonomous driving system producers must have $1 million in insurance per car, per accident, and $5 million in liability coverage. Bulgaria has a mandatory requirement for third-party liability insurance. In Italy, civil liability cover for AV must be equal to four times the cover for a non-AV vehicle. Wherever law does not specifically call out AV insurance there the general automobile insurance guidance is followed. Hence it is amply clear that any country permitting AVs to ply must revisit their vehicle insurance laws and have the requirement for the right insurance made mandatory. This will ensure the right compensation in case of a mishap and also high insurance will act as a deterrent and AV companies will be careful before launching any AV.

4.2.6 *Infrastructure for plying AVs like mandatory road markings, interactive signs, and signals*

Proper infrastructure support has, is and will always be a necessity for AVs to ply properly and safely. In April 2016 the Joint Legislative Committee ("Committee") was established in Alabama to carry out a study on AVs.[579] The Committee observed that for efficient and safe working of AVs "some amount of infrastructure[580] like satellites, road striping, interactive signs and traffic signals are mandatory to help vehicles figure out when and where to adjust the vehicle speed and how to change lanes or allow other vehicles to change lanes" and accordingly, the law was recommended to be tweaked. It was also mentioned that self-driving vehicle laws must protect not only road users but also the freight sector. Vide Bill no 4059 passed in March 2018, Oregon[581] a

---

[578.] Insurance law of PRC [online] available at https://www.ilo.org/dyn/natlex/docs/ELECTRONIC/85811/96281/F307694451/CHN85811.pdf (last visited March 25, 2021).

[579.] William Thornton | wthornton@al.com, Alabama Legislature preparing for self-driving cars AL (2019), https://www.al.com/news/anniston-gadsden/2019/03/how-the-alabama-legislature-is-preparing-for-self-driving-cars.html (last visited December 17, 2020).

[580.] Id.

[581.] Legislative Assembly of Oregon House Bill No 4059, https://legiscan.com/OR/text/HB4059/2018 (last visited December 28, 2020).

task force was established to make recommendations on infrastructure designs for facilitating AVs to run on the roads of Oregon.

Analysis of Infrastructure requirements for AVs in most jurisdictions reveals that the right infrastructure support is always a necessity for AVs to ply properly and safely and many countries across the world which are in favour of unmanned AVs are working towards the same. Similarly, there are quite a few states in the USA and many countries across the world, that mandate presence of a driver in an AV, and there are many which do not mandate it and permit remote drivers. Some of the countries which mandate the physical presence of drivers are Germany, Norway, Portugal, Russia, China, and Italy. Based on the topography of a country and its driving culture each nation will have to adopt the requirement/absence of a driver in self-driving AVs. There cannot be a one size fits all approach to this issue.

4.2.7   AV Driver (License and Physical Presence)

There are quite a few states in the USA and many countries across the world, that mandate presence of a driver in an AV, and there are many which do not mandate it.

4.2.7.1   Remote Drivers

The July 2017 Georgia bill permitted automated vehicles to be driven without a human in it[582]. Michigan in December 2016 passed three more bills related to AVs. Section 665 (5) (Bill 996)[583] states that to comply with applicable motor vehicle rules, an automated driving system that is allowed to operate without a driver must be treated as a driver when in use. Vide HB 2205 of Texas it was specified[584] that the system shall be considered as licensed to operate the vehicle[585].

[582.] State of Georgia Senate Bill No 219, https://www.legis.ga.gov/Legislation/20172018/170801.pdf (last visited December 24, 2020).

[583.] State of Michigan enrolled Senate Bill No 995, https://www.legislature.mi.gov/documents/2015-2016/publicact/pdf/2016-PA-0332.pdf (last visited December 26, 2020).

[584.] State of Texas SB 2205, https://capitol.texas.gov/tlodocs/85R/billtext/pdf/SB02205F.pdf#navpanes=0 (last visited December 28, 2020).

[585.] Id.

4.2.7.2 *Physical Drivers*

Florida passed two laws in July 2016 allowing the use of autonomous vehicles on public highways when a driver's license is in good standing. A few months before in April 2016, Virginia General Assembly enacted a law allowing an AV's safety driver to watch a factory-installed video display while the AV is moving[586]. All AV testing in Italy is supposed to include a safety driver in the car who can take over the AV if necessary. Similarly, as per the initial Norwegian regulations every AV is mandated to have an employee on board to apply emergency brakes if required. The Road Traffic Code of Portugal requires all vehicles to have drivers with valid legal driving licenses. The driver of the AV in Russia must possess a clean driving record and at least three years of experience. Additionally, the driver must be deemed medically suitable to serve as a driving teacher for the type of vehicle being evaluated[587]. ICV testing driver in PRC or China needs to possess certain prescribed personal qualifications.

Analysis of the legal regime in the countries/states which permit driverless operations and countries which mandate the presence of a physical driver in a self-driving car indicates that the jurisprudence in arriving at that decision is mainly based on the contours of the country/state roads, existing driving habits, existing driving speeds and kind of infrastructure available for AVs in those countries/states.

4.2.8 *Product Liability*

Product liability means the liability of any or all parties involved in the manufacturing cycle for any kind of damage caused by that product. Product liability cases may be based on strict liability, carelessness, or a breach of the fitness warranty. In strict liability cases, the intent of the defendant is no excuse. Any person may be held liable under strict liability, but only businesses with a commercial goal may be held liable under absolute liability. In absolute liability, an organisation can be held accountable even in the absence of a harmful thing

---

[586.] Josh Mandell @joshuamandell et al., Virginia still a blank slate for self-driving car laws Charlottesville Tomorrow â€¢ Informed citizens create better communities, https://www.cvilletomorrow.org/articles/virginia-blank-slate-for-self-driving-car-laws (last visited February 20, 2021).

[587.] Government.ru. 2021. *О проведении эксперимента по эксплуатации на автодорогах высокоавтоматизированных транспортных средств.* [online] Available at: <http://government.ru/docs/34831/> [last visited June 28, 2021].

escaping; in strict liability, this is not the case. Product liability laws for AVs are on different scales across the world. In the case of AVs, product liability can be for both hardware and software. As regards product liability, the manufacturer and/or the AV software developer would be subject to existing general strict liability rules for damages caused by the AV in "autonomous mode". The only exception would be if it is proved that the damage was unavoidable. In any case, liability will only arise when the damages are caused by a faulty part or solution applied by the developer.

As per State of Nevada Assembly Bill No 69, the person, as described in Section 470, who builds a vehicle and installs autonomous technology on it once it is initially completed is a manufacturer of autonomous technology, or, if a vehicle was not built with autonomous technology by the car maker, the person who modifies the vehicle after it has been built by adding autonomous technology to make it an autonomous vehicle is the manufacturer. Michigan vide Senate bill 663[588] of December 2013 and Senate bill 169 of March 2014 has permitted testing of automated vehicles under laid down conditions[589]. Under Sec 817, the Bill also provided immunity from civil liability to the manufacturer of automated technology for damages caused due to modifications made by a third party to a motor vehicle/automated motor vehicle. When repairs are made to automated vehicles per manufacturer requirements, the bill also shields auto mechanics and repair facilities from product responsibility.

As per Austrian laws product liability also applies to software that's part of a physical product like an AV. Any bug or error in the software which results in an accident by an AV shall lead to product liability of the car manufacturer[590]. As regards AI software sometimes errors occur basis what the sensors decode. Things are still not very clear on how these situations will be handled especially from a product liability perspective. Imposing product liability on a manufacturer for this may not be the right approach, but the fact is every AV manufacturer carries this risk. As per Austrian law, warranty claims can only be made against the seller and not against the manufacturer

---

[588.] State of Michigan Enrolled Senate Bill No 663, https://www.legislature.mi.gov/documents/2013-2014/publicact/pdf/2013-PA-0251.pdf (last visited December 26, 2020).
[589.] State of Michigan Enrolled Senate Bill No 169, https://www.legislature.mi.gov/documents/2013-2014/publicact/pdf/2013-PA-0231.pdf (last visited December 26, 2020).
[590.] Autonomous Vehicles Law and Regulations in Austria [online] available at https://cms.law/en/int/expert-guides/cms-expert-guide-to-autonomous-vehicles-avs/austria (last visited March 2, 2021).

of the AV[591] and car manufacturers will be subject to strict liability under the applicable product liability laws. The maker of a product, its representatives, and in some circumstances the seller must assume responsibility to assure the safety and compliance of the product, according to the Bulgarian Consumer Protection Act of 2005.

In Italy, the general liability rules would apply to manufacturers for defects in the vehicle, including the mandatory two-year consumer warranty for defects. Monaco is in the process of drafting a legislative proposal for a special system of liability for defective products to be borne by producers/manufacturers if the product is the cause of personal injury or material damage. In Poland, the manufacturer can be held liable for product liability under the general liability laws and could be made liable under liability damages for hazardous products, which are moveable goods and carry strict liability regardless of fault[592].

In Portugal, the general liability rules of Portuguese law would be applicable for product liability, and the manufacturer could be held liable for product damage including damages caused due to defect in the AV[593]. According to Romanian civil law regulations, the manufacturer is liable for product liability in instances when flaws cause death, physical harm, or damage to goods. In addition, the manufacturer can also be made criminally liable for product defects as per the existing criminal law of Romania. In line with what exists for the manufacturer, the AV system provider can also be held liable for civil and criminal negligence regarding damages caused by the operating system. As per consumer laws in Russia, consumers can file general liability claims for the warranty of AVs, against the manufacturer. Thus, an AV manufacturer will face similar liability to a regular car manufacturer once they start selling AVs to consumers[594]. In Turkey, for damages caused due to defects in the vehicle, the manufacturer could be held liable under the Consumer Protection Law and asked to provide a free repair of the vehicle or replace the vehicle.

[591.] Id.

[592.] Autonomous Vehicles Law and Regulations in Poland [online] available at https://cms.law/en/int/expert-guides/cms-expert-guide-to-autonomous-vehicles-avs/poland (last visited March 2, 2021)

[593.] Autonomous Vehicles Law and Regulations in Portugal [online] available at https://cms.law/en/int/expert-guides/cms-expert-guide-to-autonomous-vehicles-avs/portugal (last visited March 2, 2021).

[594.] CEE Legal Matters. 2021. *Autonomous Vehicles Regulation in Russia.* [online] Available at: <https://ceelegalmatters.com/cms/13891-autonomous-vehicles-regulation-in-russia> [Last visited June 28, 2021].

Since Slovenia has no special rules concerning liability for damages caused by an AV, the general rules of liability would apply to the user and manufacturer. Rules of strict liability apply to the manufacturer and the AV software developer for any defect in the vehicle or software which results in damages[595]. As per Slovenian laws, warranty claims can be made against the seller and manufacturer of the vehicle. If the vehicle had any manufacturing defect, then that's not covered by warranty and the manufacturer will be subject to strict liability under the applicable rules of the Obligations Code[596]. For product defects in Ukraine in the case of AVs, as per the general rules of defect of products claims, the manufacturer can be held liable. There is no limitation on the manufacturer's liability under the Commercial Code of Ukraine[597]. If the manufacturing defects are of a serious nature defect, a customer may either terminate the sale purchase agreement and seek a refund or demand substitution of the vehicle.

Concerning manufacturer's liability for ICVs, present PRC law has no specific provision and hence warranty claims will have to be solved either on basis of the provisions of the sales contract or under the PRC Product Quality Law. On December 29, 2012, the PRC State Quality Supervision, Inspection, and Quarantine Bureau promulgated the Automobile Guarantee Provisions as per which any seller who sells consumer automobiles will mandatorily assume three guarantee liabilities towards their consumers[598]. These three guarantee liabilities[599] are guarantee liability for repair, replacement, and return. There is one more law in PRC for recalling defective vehicles. All manufacturers are obliged to recall sold vehicles if any defect is later found in them which needs attention at the manufacturers level. This liability is not contractual and is a statutory obligation imposed by law[600].

---

[595]. Autonomous Vehicles Law and Regulations in Slovenia [online] available at https://cms.law/en/int/expert-guides/cms-expert-guide-to-autonomous-vehicles-avs/slovenia (last visited March 2, 2021).

[596]. Id.

[597]. Autonomous vehicles law and regulation in Ukraine CMS Expert Guides, https://cms.law/en/int/expert-guides/cms-expert-guide-to-autonomous-vehicles-avs/ukraine (last visited April 2, 2021)

[598]. Automobile Guarantee regulations for Automobiles used by Consumers [online] available at https://cms.law/en/chn/publication/automobile-guarantee-regulations-for-automobiles-used-by-consumers-have-finally-been-promulgated (last visited March 25, 2021).

[599]. Id.

[600]. Autonomous vehicles law and regulation in China, CMS Expert Guides, https://cms.law/en/int/expert-guides/cms-expert-guide-to-autonomous-vehicles-avs/china (last visited April 29, 2021).

The 2017 Singapore Rules do not specifically prescribe details of manufacturers' responsibility for an AV being driven in an automated mode. However, the manufacturer's current obligations will remain in force[601]. As regards warranty the current broad consumer protection or product liability laws would be made applicable by the courts. In Vietnam, as regards, product recalls, as per Circular No. 30/2011/TT-BGTVT, products failing to satisfy any applicable technical requirement, or which are likely to endanger humans or property due to such technical errors have to be recalled by the manufacturer within five days of detecting the error[602]. In such a case the manufacturer is also required to submit a written notification to the Vietnam Register with a proposed recall plan. As regards product liability Vietnam's Civil Code and the Law on Consumer Rights Protection prescribe the right guidance to calculate the liability for damages.

Analysis of product liability laws in most jurisdictions reveals that they all have one thing in common which is the protection of the customer. Concerning product liability most states in the USA which permit AVs to ply have well-defined product liability laws in place. Most of these rules shield car technicians and repair shops from product responsibility when they work on automated vehicles under manufacturer guidelines. This is very much essential else repair facilities will not come up and if there is a breakdown it can become a nightmare. The absence of proper repair facilities can indirectly hamper the growth and adoption of AVs. Law relating to recalling defective vehicles exists in many jurisdictions and China, the guarantee liabilities come in three buckets, and they are guaranteed liability for repair, replacement, and return. Immunity is also generally provided from civil liability to the manufacturer of automated technology for damages caused due to modifications made by a third party to a motor vehicle/automated motor vehicle. This aspect of the law is also very much essential as manufacturers cannot be made responsible for damages caused by third parties who did not follow the manufacturer's specifications. In most jurisdictions, product liability also applies to software that's part of a physical product like an AV. Wherever specific product liability rules are not specified the general liability rules would apply to the manufacturers.

[601.] Autonomous vehicles law and regulation in Singapore, CMS, https://cms.law/en/int/expert-guides/cms-expert-guide-to-autonomous-vehicles-avs/singapore (last visited April 30, 2021).

[602.] Ministry of Transport [online] available at https://vanbanphapluat.co/circular-no-30-2011-tt-bgtvt-on-the-environment-protection-and-technical-safety (last visited March 24, 2021).

A recent example of a product recall, relating to AVs is the Cruise vehicle crash which occurred in California, resulting in minor injuries. A cruise car entered a left-turn lane on June 3, 2022, signalled, and then began making a left turn as soon as the light turned green. At the same time, a 2016 Toyota Prius was approaching the intersection from the opposite direction in the right-turn lane, travelling at around 40 mph in a 25-mph zone. The approaching Prius reached the intersection as the Cruise automatic vehicle was making a turn, slamming into its rear passenger side.[603]. Following the tragedy, Cruise looked into what caused it and decided to Suo-moto recall and update the software in 80 of its cars[604].

4.2.9   *Vehicle Data.*

As of date, AVs with lower levels of autonomy generate around 25Gb data per hour and with increasing features coming soon this number will grow and managing the whole process is going to become extremely complex. AVs generate personal data, sensor data, location data and v2x or vehicle-to-everything data. A new EU rule would require black boxes to be placed in all new cars starting in 2024. The safety of data and concerns for data privacy of such generated data are other important aspects lawmakers across the world consider when making laws for handling AV data. In April 2016 the Joint Legislative Committee ("Committee") was established in Alabama to carry out a study on AVs.[605] The Committee recommended that Vehicles must also have recording devices. Bill HB 101 of Utah talks of vehicle data being private data of the vehicle owner[606]. Alabama Joint Legislative Committee [607] in April

---

[603] Business World, US Agency Probing Self-Driving Cruise Car Crash in California, https://www.businessworld.in/article/U-S-Agency-Probing-Self-Driving-Cruise-Car-Crash-In-California/08-07-2022-436208/ (last visited July 17, 2022).

[604] The San Francisco Standard, San Francisco Cruise Crash Leads to Software Recall in 80 Self-Driving Vehicles, https://sfstandard.com/business/san-francisco-cruise-crash-leads-to-recall-of-80-self-driving-vehicles/ (last visited July 17, 2022).

[605] William Thornton | wthornton@al.com, Alabama Legislature preparing for self-driving cars AL (2019), https://www.al.com/news/anniston-gadsden/2019/03/how-the-alabama-legislature-is-preparing-for-self-driving-cars.html (last visited December 17, 2020).

[606] Betsy Foresman, Utah to join states allowing fully autonomous vehicles on public roads StateScoop (2019), https://statescoop.com/utah-to-join-states-allowing-fully-autonomous-vehicles-on-public-roads/ (last visited December 28, 2020).

[607] William Thornton | wthornton@al.com, Alabama Legislature preparing for self-driving cars AL (2019), https://www.al.com/news/anniston-gadsden/2019/03/how-the-alabama-legislature-is-preparing-for-self-driving-cars.html (last visited December 17, 2020).

2016 mandated that AVs must also be equipped with recording devices and carry insurance.

The StVG of Germany also has provisions concerning data and handling of data (§63a). The provisions concerning data storage and transmission are enumerated in Article §63a (4) of the StVG. The ask in those provisions are that the AV owner is mandated to delete the data, which is stored after six months, whenever sought by law enforcement authorities the stored data can be forwarded to those authorities and lastly in cases where data is forwarded to law enforcement authorities at their request then the owner of the AV must delete the stored data (and transmitted) after three years. As per Russian regulations, every AV must mandatorily have a data recording system and the driver must also be able to switch off and on the equipment for traffic recording and actions of the driver[608]. The Automated Lane Keeping System or ALKS[609] of the UNECE (United Nations Economic Commission for Europe) has a requirement to have a Blackbox in all AVs which can record every minute detailed recording of the AV's journey and the way the AV was handled during the journey or an accident.

As regards data privacy for AVs in Indonesia, Minister of Communication and Informatics (MOCI) Regulation No. 20 of 2016 (DP Regulation) is equally applicable to data collected by AVs or AV developers since Indonesia at present does not have a personal data protection law covering a broader range of subjects. The Data Protection (DP) Regulation prescribes that any overseas transfer of personal data must first receive MOCI permission[610]. Hence under normal circumstances, AV developers would be required to store data in Indonesia. With proper permission, AV developers can store the biometrics data offshore. However, before such storing, as per MOCI 20/2016 AV developer must give notice of such offshore transfer to the data owners and seek their consent[611] before transferring such data. In case of a breach, AV

[608.] Government.ru. 2021. О проведении эксперимента по эксплуатации на автодорогах высокоавтоматизированных транспортных средств. [online] Available at: <http://government.ru/docs/34831/> [Last visited June 28, 2021]. The mandatory preservation period of this video is ten years, and it must be mandatorily provided to any government agency when requested.

[609.] Compliance with new Automatic Lane Keeping System Regulation, ALKS, Supra Note 498, at 136.

[610.] Data Protection in Indonesia: Processing requirements [online] available at https://www.ssek.com/blog/data-protection-in-indonesia-processing-requirements (last visited March 27, 2022). MOCI Regulation 20 specifically stipulates that the consent must be in writing and can be provided manually or electronically.

[611.] Id.

developers must notify the data owners in writing within 14 days after the incident.

Analysis of the various data protection regimes globally specifically concerning AVs reveals that the importance of data generated by AVs is well known and most jurisdictions have laws to protect the same. The requirement of recording devices and processes for handling and storage of data including deletion is spelt out in many laws. Some of the notable ones are Alabama, Utah, Germany, and Russia. As per Russian regulations, every AV must mandatorily have a data recording system and the driver must also be able to switch off and on the equipment for traffic recording and actions of the driver. The Automated Lane Keeping System or ALKS of the UNECE has a requirement to have a Blackbox in all AVs which can record every minute detailed recording of the AV's journey and the way the AV was handled during the journey or an accident. In Indonesia, the Data Protection Regulation prescribes that any overseas transfer of personal data must first receive proper authorization. Thus, it is amply clear that since AVs generate a huge amount of data, any AV law without specific provisions on data protection and handling would make it an incomplete law.

4.2.10 SAE levels of Automation

Autonomous vehicles are rated according to their automation levels. The minimum standard for automation is six levels. The society of Automotive Engineers in the USA (SAE) has designed a zero to five rating system to categorise different levels of automation. Greater automation means a higher level and lower automation means a lower level[612]. The various levels are:

- **Level Zero: No Automation** – The driver is entirely in charge.

- **Level One: Driver Assistance** – The automobile has been programmed to do partial work like say cruise control. However, the driver has to be present and must perform assigned tasks.

- **Level Two: Partial Automation** – Help is provided to the driver by the vehicle on one or more systems. Driver's presence is mandatory.

---

[612] SAE Levels of Driving Automation [online] available at https://www.sae.org/blog/sae-j3016-update (last visited, February 22, 2022)

- **Level Three: Conditional Automation** – AV is automated in all aspects, but the presence and intervention of the driver are required on an as-needed basis.

- **Level Four: High Automation** – The automobile is fully automated and can perform all duties even if driver intervention is not there. However, the driver's presence is mandatory.

- **Level Five: Full Automation** – Fully automated driverless vehicle.

Wherever AVs are allowed to be tested in the USA normally there are no restrictions on the levels subject to safety requirements being maintained. However, outside the USA, the situation is different. Level 5 testing is not permitted in Poland as the rules in Poland mandate the requirement of a safety driver to take control. Level 4 testing is permitted subject to compliance with laid down procedures and the availability of a safety driver at all times to take control if required. Level 3 again is permitted for testing almost on similar lines to Level 4. As of date consumers are not allowed to ply AVs on the roads in Poland[613]. Level 3-5 AVs are not allowed on Portugal roads except in line with various European projects[614]. As per existing Russian regulations, consumers are not allowed to use AVs with SAE levels 3 to 5 on their public roads[615]. In Slovenia, consumers are permitted to use AVs with features only up to SAE Level 2 (e.g., Autonomous Parking). Driving AVs (SAE Levels 3-5) is not permitted on public roads in Switzerland. However, conditional testing can be permitted.

As per general regulation for Ukrainian traffic rules (approved by the Regulation of the Cabinet of Ministers dated 10 October 2001 No. 1306), a human driver is mandatory to operate a vehicle[616]. The driver must be always attentive and never be distracted from driving. Because of the foregoing use of AVs (SAE levels 3 – 5) on public roads in Ukraine would be non-compliant with Ukrainian traffic rules. As regards motor vehicle laws there are no officially

---

[613.] Autonomous Vehicles Law and Regulations in Poland [online] available at https://cms.law/en/int/expert-guides/cms-expert-guide-to-autonomous-vehicles-avs/poland (last visited March 2, 2021)

[614.] Autonomous Vehicles Law and Regulations in Portugal [online] available at https://cms.law/en/int/expert-guides/cms-expert-guide-to-autonomous-vehicles-avs/portugal (last visited March 2, 2021).

[615.] Government.ru. 2021. *О проведении эксперимента по эксплуатации на автодорогах высокоавтоматизированных транспортных средств.* [online] Available at: <http://government.ru/docs/34831/> [Last visited June 28, 2021].

[616.] Autonomous vehicles law and regulation in Ukraine CMS Expert Guides, https://cms.law/en/int/expert-guides/cms-expert-guide-to-autonomous-vehicles-avs/ukraine (last visited April 2, 2021)

released standards in China (PRC) akin to the standards of the Society of Automotive Engineers (**SAE Levels**). The Chinese government on 09 March 2020 released the draft National Standards for AVs[617]. This Standard classifies AV driving on a pattern just like SAE levels.

As of now Japan used SAE level 3 AVs[618] in the Olympic games 2020 (which was moved to 2021). It is expected that Japan will roll out SAE level 4 vehicles[619] for testing and the final commercial offer by 2025. Additionally, there are plans to conduct tests in the shipping sector or remote, less populous places. The Japanese government's primary goal is to implement level 3 first and use it as an advantage to touch level 4. Concerning the Japanese Motor Vehicle Act, to date, the said act was designed to only handle AVs upto SAE level 3. With the latest revisions to the Act i.e., Law No 14 and 20 enacted in 2019 the act now has adequate provisions to handle level 4 SAE AVs also. On the tenth of December 2018, South Korea opened the K-City test facility to test and commercialize AVs especially, Level 3 AVs which would also include sophisticated features but still require a safety driver to take control of the automobile as and when required[620]. In comparison with other countries which are pioneers in the AV sector, South Korean legislation only permits the driving of level 3 AVs due to its slower legislation process.

Analysis of SAE rules and their adoption globally reveals that Level 1 and 2 is allowed in most countries. Level 3-5 is restricted and allowed only in a few countries across the world. Russia, Slovenia, and Ukraine, as of date do not allow Level 3 to Level 5 on their public roads. China has standards akin to SAE levels. Quite a few states in the USA allow upto Level 5 for testing and even use. However, one thing is of paramount importance here and that is the grading of vehicles basis SAE or equivalent levels. This is an important aspect or requirement for any legal regime permitting AVs to ply and based on the existing infrastructure development a country/state may legally permit AVs upto a particular SAE level only for operation.

---

[617] Global guide to Autonomous Vehicles 2021 [online] available at https://www.businessgoing.digital/global-guide-to-autonomous-vehicles-2021/ (last visited, March 25, 2021)

[618] Outline of systematic preparations related to autonomous driving (2018), https://www.kantei.go.jp/jp/singi/it2/kettei/pdf/20180413/auto_drive.pdf, p.6 (last visited April 30, 2021).

[619] ITS 2019 (2019), https://www.kantei.go.jp/jp/singi/it2/kettei/pdf/20190607/siryou9.pdf, p.21 (last visited April 30, 2021).

[620] South Korea builds K-City for AV testing [online] available at https://en.yna.co.kr/view/AEN20181210008800320 (last visited March 25, 2021).

4.2.11 Penalty regime for violation in various jurisdictions

We have seen the product liability for AVs in various jurisdictions and now looking at the Criminal liability for accidents mainly it can be the driver when there is one in the driver's seat or the organisation which oversees a vehicle without a driver or the ASDE, or authorised self-driving entity, which is the software or vehicle maker that submits the car for authorization, or a joint venture between the two. This is the entity which must respond to regulatory action and oversee any recall activity. Different jurisdictions across the world have afforded different treatment to this complex aspect of AV accident liability.

4.2.11.1 Liability as per existing laws

As per Colorado Senate Bill 17-213, Liability for any driverless AV accident shall be based on the relevant State, Federal, and Common Law. Section 4 of Nevada SB 313[621] requires AVs to meet all applicable federal standards and regulations before being permitted to register in the state and any violation can result in fines upto $2500. In Austria, the country follows the general liability rules for any damage caused by an AV or semi-AV. The general Austrian law clearly distinguishes between liability for damages due to negligent behaviour and strict (product) liability of the car operator or the car manufacturer according to the product liability law[622]. If any driver violates any rules the driver will be liable[623] for all damages. As regards damages to third parties or pedestrians normally the insurance will cover it. Since AVs are not permitted in Bulgaria the Road Traffic Act is silent on liability for AV accidents. For conventional cars, the driver and sometimes the owner can be held liable for damages.

Germany has no specific rules for liability on account of accidents caused by AVs[624] and hence the general liability law will determine the liability of the owner. Such liability can also include criminal liability. The same is the

---

[621.] State of Nevada Senator Bill No 313, https://www.leg.state.nv.us/Session/77th2013/Bills/SB/SB313_EN.pdf (last visited December 26, 2020). n

[622.] Autonomous Vehicles Law and Regulations in Austria [online] available at https://cms.law/en/int/expert-guides/cms-expert-guide-to-autonomous-vehicles-avs/austria (last visited March 2, 2021).

[623.] Id.

[624.] German Civil Code [online] available at https://www.gesetze-im-internet.de/englisch_bgb/ (last visited March 21, 2021).

case with Italy[625], Monaco and Poland. In case of any liability issues in Italy, under normal circumstances, the owner and driver would be held jointly and severally responsible unless the owner can demonstrate that the car was operated against his will. Any liability claim in Monaco would be dealt with under the Monegasque Civil Code[626] and in Poland under general liability laws. Exemption from liability in Poland will kick in only during force majeure or in a case where the fault lies with the person who suffered the damage.

Portugal and Romania have no special AV laws and hence Liability as of date would be determined based on the general principles of civil liability in these countries. Any damage brought on by the car would be the responsibility of the owner/driver. A Portuguese owner may still be liable, even when the owner is not the driver if the vehicle is driven under his direction and in his interests[627]. As per the existing Romanian liability regime[628], the liability of the owner is subject to civil liability as per the existing civil law and criminal liability as per existing criminal law. Exceptions granted would be for reasons of force majeure and contributory negligence. Swiss laws handle liability in three buckets viz Owner, Operator and Manufacturer[629]. As regards AVs there is no specific liability-related legislation as of the date and hence the general liability rules would apply for damages caused by an AV.

Turkey has no separate AV laws and hence the same strict liability principles would apply while handling liability cases of AVs. The Turkish Code of Obligations and the Consumer Protection Law would be applicable. Under the existing Turkish criminal law, only natural persons can be held criminally liable and not legal persons like manufacturers. Hence programmers of the AVs, drivers and users of the vehicles could be held criminally liable if they negligently or intentionally commit criminal offences (taksirli suç ve kasti suç)[630]. Similarly, since Ukraine also has no AV-specific laws, as per the Civil

---

[625] Autonomous Vehicles Law and Regulations in Italy [online] available at https://cms.law/en/int/expert-guides/cms-expert-guide-to-autonomous-vehicles-avs/italy (last visited March 2, 2021)

[626] Autonomous Vehicles Law and Regulations in Monaco [online] available at https://cms.law/en/int/expert-guides/cms-expert-guide-to-autonomous-vehicles-avs/monaco (last visited March 2, 2021).

[627] Autonomous Vehicles Law and Regulations in Portugal [online] available at https://cms.law/en/int/expert-guides/cms expert guide to autonomous vehicles avs/portugal (last visited March 2, 2021).

[628] Autonomous Vehicles Law and Regulations in Romania [online] available at https://cms.law/en/int/expert-guides/cms-expert-guide-to-autonomous-vehicles-avs/romania (last visited March 2, 2021).

[629] Autonomous Vehicles Law and Regulations in Switzerland [online] available at https://cms.law/en/int/expert-guides/cms-expert-guide-to-autonomous-vehicles-avs/switzerland (last visited March 2, 2021).

[630] Autonomous vehicles law and regulation in Turkey CMS Expert Guides, https://cms.law/en/int/expert-guides/cms-expert-guide-to-autonomous-vehicles-avs/turkey (last visited March 22, 2021)

Code of Ukraine, or as per the Law of Ukraine for any liability on account of damages attributable to product defect, the AV's manufacturer or recorded importer may be held accountable for the defective product and any harm it causes.

4.2.11.2 Liability of manufacturer

Section 4 of DC Act 19-643 of 2013[631], prescribes that "the original manufacturer of a vehicle shall not be liable for any accident caused by a vehicle which was converted as an autonomous vehicle by a third-party vendor". Also, as per Section 3, a human driver must be there and prepared to take over in an autonomous vehicle, if necessary[632].

4.2.11.3 Liability of AV as a driver

As per the AV Act of Tennessee (SB 151 of 2017), Automated Vehicle is specified as a driver in high or full automation mode and hence any offence by such AVs shall mean a charge on the AV as per existing laws. This is the only jurisdiction in the world where an AV can be charged. How practical this law is, has to be seen only when a case comes up.

4.2.11.4 Liability of Operator/Driver/Owner

Austrian law also provides for strict liability of the operator of any car irrespective of the fact that the car is autonomous or not. As per Austrian law, any damage caused by the vehicle would be the responsibility of the driver, unless the car was used without the operator's consent. The operator as per Austrian law is jointly and severally liable for failure to apply due care in preventing misuse of the car without consent (e.g., a mother does not lock her car and her child secretly drives away and hits a pedestrian). As per Russian regulations, if no one is found guilty of traffic and other accidents on Russian roads, the AV owner will be liable. In Russia, conventional civil law may be used to handle AV accident complaints and there is a fair possibility that the owners of such driverless vehicles can be made financially liable irrespective of the damage severity. However, there are quite a few in Russia who also have opined that such liability can be placed on the AV drivers who fail to monitor

---

[631]. DC Act 19-643 [online] available at https://lims.dccouncil.us/downloads/LIMS/26687/Signed_Act/B19-0931-SignedAct.pdf (last visited February 23, 2022).
[632]. Id.

the AV systems or on developers of faulty AV technology, especially in cases of system failure[633]. The driver of a two-track vehicle must constantly maintain steering control under the Slovenian Road Traffic Rules Act, or else they risk receiving a 300 Euro fine. The permits granted for testing in Sweden stipulate that in the case of fully Autonomous Vehicles, the legal responsibility for the operation of AVs and the resultant damages caused by such AVs shall be on the party which obtained the testing permit. However, in the case of a vehicle operating at lower automation levels and having a driver control the vehicle the responsibility for bearing civil and criminal liability shall be on the driver.

In PRC an Entity carrying out ICV testing must have sufficient financial capability to pay up any kind of compensation on account of damages to people or property brought on by ICV testing. In case of any accident during the testing of ICV compensation for damages shall be determined as per existing laws. Some of the broad guidelines/rules[634] as per existing PRC laws state that for accidents between two motor vehicles is that the party at fault must bear the liability, and in cases when the fault is shared by both parties, the assessment of culpability shall be based on the fault in proportion. Under the 2017 Rules of Singapore, the AV owner has to mandatorily have a liability insurance policy for the vehicle and if the owner is not able to do that, then a security deposit must be placed with the LTA to compensate for any payments due on account of any death, bodily injury or damage to property[635]. The liability of the driver or operator would be decided as per current laws.

Analysis of liability regimes for AVs globally reveals that accidents caused by AVs are complex to dissect when it comes to liability and different jurisdictions across the world have afforded different treatment to this complex aspect of AV accident liability. It is required by law to identify the person who was using the AV to commit the crime. It appears that this will be decided by taking into account the unique circumstances of each alleged act[636]. These actors may affect the AV by controlling it remotely, providing

---

[633]. Autonomous Vehicles Law and Regulations in Russia [online] available at https://cms.law/en/int/expert-guides/cms-expert-guide-to-autonomous-vehicles-avs/russia (last visited March 2, 2021).

[634]. 2019 Autonomous Vehicles Readiness Index (2019), https://assets.kpmg/content/dam/kpmg/xx/pdf/2019/02/2019-autonomous-vehicles-readiness-index.pdf (last visited April 29, 2021).

[635]. Road traffic Autonomous Motor Vehicles Rules 2017 [online] available at https://sso.agc.gov.sg/SL/RTA1961-S464-2017?DocDate=20170823 (last visited March 25, 2021).

[636]. Artificial Intelligence and the External element of the Crime [online] available at http://www.diva-portal.org/smash/get/diva2:1115160/Fulltext01.pdf (last visited February 19, 2022)

precise instructions, or failing to step in and overturn the AV's decisions. If the AV is viewed as a straightforward machine controlled by a human, the first of the scenarios is not very challenging to resolve from the standpoint of culpability[637]. But as soon as the AV behaves autonomously, disregards the user's instructions, and causes significant harm, the issue of culpability becomes more challenging. Is it the user's or the supervisor's duty to override the AV's improper behaviour?[638]

The manufacturer is primarily accountable for the AV's hardware and software, which covers everything from mechanical components to internal code and algorithms, as well as the AV's education and training[639]. Since the AV's code serves as its brain, as well as its core and the key to all it is capable of, the manufacturer may have an impact on the AV in any area. The manufacturer and its agents will likely be held accountable for any defect that results from a manufacturer's negligence[640]. But what if the AI unintentionally commits errors that the manufacturers and programmers were unaware of when they built the code? Do these issues need to be resolved by the user or the supervisor? Although the owner and user or supervisor primarily overlap, it is best to rule out the owner as a prospective defendant at this stage of the analysis even though they do. In reality, one of the key players in the future of liability will be the owner[641].

Criminal responsibility is clear for any outsider who in some way affects the AV's decision-making. The external party could be a hacker with the capability to remotely control the AV or changes its code and thereby ensure that the AV works in a certain way[642]. The hacker may also be a third party who approaches the AV with a suggestion. One may even consider an outsider deceiving the AV into misinterpreting its surroundings in a way that leads to wrongdoing[643]. Colorado, Nevada, Austria, Bulgaria, Germany, Monaco, Poland, Portugal, Switzerland, Turkey, Singapore and Ukraine attribute liability based on existing laws as they do not have separate laws for AVs. As per Tennessee laws, if an AV is operating in full automation mode, then that

---

637. Id.
638. Id.
639. Id.
640. Id.
641. Id.
642. Id.
643. Id.

AV will be considered a driver for any offence. In Austria, Russia and Slovenia the driver/operator/owner can be made liable for any AV accidents. In PRC an Entity carrying out ICV testing must have sufficient financial capability to pay up any kind of compensation due to personal or property damages caused by ICV testing. Liability for any kind of accident or incident which concerns self-driving AVs is part of the legal regime which is still under development and there is a need for the current liability laws to be modified to come up with the right kind of remedy for any self-driving AV-caused accident/damage.

4.2.12 *Overview of various AV testing committees and Government bodies across various Jurisdictions for AV implementation*

To move ahead in the development of AVs during 2017 The Safely Ensuring Lives Future Deployment and Research in Vehicle Evolution Act or Self-Drive Act was approved by the US House of Representatives vide H.R. 3388[644]. This bill was primarily aimed at adopting federal safety standards in the arena of AVs. Post passing of this Act, a similar bill called American Vision for Safer Transportation through Advancement of Revolutionary Technologies Act (AV START Act) under reference S. 1885 was also presented, but never passed[645]. The AV Start Act, which is very similar to the proposed language of the SELF-DRIVE Act, calls for the creation of a Highly Automated Vehicle Advisory Council within the Department of Transportation to assess self-driving car-related issues. It also addresses federal regulation of autonomous vehicle testing[646].

Under the AV Start Act, the onus was vested in each AV manufacturer to provide a defined strategy for recognising and minimising cybersecurity risks. The Self-Drive Act clarified the role of NHTSA in regulating the safety, design, and construction of AVs so that NHTSA could plan any required upgrades

---

[644.] Congress Taking Another Look at Regulating Automated Driving Systems, JD Supra, https://www.jdsupra.com/legalnews/congress-taking-another-look-at-51054/ (last visited December 17, 2020).

[645.] S 1885 – AV Start Act [online available at https://www.congress.gov/bill/115th-congress/senate-bill/1885#:~:text=This%20bill%3A%20(1)%20establishes,report%20subject%20areas%3B%20(3) (last visited February 23, 2022).

[646.] Id.

and research and development (R&D)[647]. The Self-Drive Act created NHTSA safety evaluation certifications for manufacturers and improved protection for consumer education, cybersecurity, and privacy.

Maine in January 2018 passed executive order and created a diverse Highly Automated Vehicles (HAV) Advisory Committee[648], having at least 11 members, including the Secretary of State, the Bureau of Insurance, the Office of Aging and Disability, and the Commissioner of the Maine Department of Transportation. Similarly, in January 2017, North Dakota[649] mandated the Department of Transports to gather data on the usage of AVs on highways and analyse them. Oregon[650] in March 2018 through a bill created a task force to coordinate programs and policies on Automated Vehicles and make a recommendation on various aspects of AVs including licensing, registration, reporting of accidents, insurance, and liability[651].

Pennsylvania does not have specific laws for AVs nor has permitted their plying on state roads. However, on 20 July 2016 vide Bill no SB 1267 the state-sanctioned a usage of upto $40 million[652] to develop technology connected with AVs. Tennessee in June 2017 created the Automated Vehicles Act[653], which was made applicable only to Automated Vehicles in high or full automation mode and had requirements for AVs regarding seat belts, child restraints, abandoned cars, and crash reporting. In June 2019 Vermont came up with the Automated Vehicles Testing Act[654]. This Act was instrumental

---

[647] H.R. 3388 (115th): Self Drive Act, GovTrack.us, https://www.govtrack.us/congress/bills/115/hr3388/summary (last visited December 17, 2020).

[648] Maine Executive Order 2018-001 [online] available at https://www.maine.gov/mdot/autonomous-vehicles/docs/2018-001_HAVExecOrder.pdf (last visited February 23, 2022).

[649] Legislative Assembly of North Dakota House Bill No 1202, https://www.legis.nd.gov/assembly/65-2017/documents/17-0711-04000.pdf (last visited December 28, 2020).

[650] Legislative Assembly of Oregon House Bill No 4059, https://legiscan.com/OR/text/HB4059/2018 (last visited December 28, 2020).

[651] Id. As per (3)(a) This section does not apply to a person operating a vehicle that is part of a connected automated braking system. (b) As used in this subsection, "connected automated braking system" means a system that uses vehicle-to-vehicle communication to electronically coordinate the braking of a lead vehicle with the braking of one or more following vehicles.

[652] Legislative Data Processing Center, 2016 Act 101 The official website for the Pennsylvania General Assembly., https://www.legis.state.pa.us/cfdocs/legis/li/uconsCheck.cfm?yr=2016&sessInd=0&act=101# (last visited December 28, 2020).

[653] Tennessee SB151, Track Bill, https://trackbill.com/bill/tennessee-senate-bill-151-motor-vehicles-as-enacted-enacts-the-automated-vehicles-act-and-other-requirements-related-to-the-operation-of-autonomous-vehicles-on-the-public-roads-of-this-state-amends-tca-title-5-title-6-title-7-title-39-title-40-title-54-title-55-title-56-title-65-and-title-67/1375094/#/details=true (last visited December 28, 2020).

[654] Vermont Laws, State House Dome, https://legislature.vermont.gov/statutes/fullchapter/23/041 (last visited December 28, 2020).

in bringing about a process for testing automated vehicles[655], which mainly required a testing permit before testing of AVs, from the Traffic Committee, which is composed of the Secretaries of Transportation, Motor Vehicles, and Public Safety[656]. The Washington Executive Order of 2018 for AVs included two main provisions regarding the formulation of a workgroup and conducting of pilot programs for AV testing[657].

The Wisconsin Governor's Steering Committee established in May 2017, was tasked with coordinating with various agencies concerning criteria for AV equipment, registration, licensing, insurance, and traffic laws. The committee was also tasked with reviewing laws that impede testing/deployment. In August 2015, the Arizona Governor issued an order that gave various agencies specific instructions to help in the testing and operation of AVs on public roads[658]. An updated order was issued in March 2018 mandating Arizona to step up towards the latest technological developments in the field of automated vehicles and ensure that all automated vehicles in the state follow the Federal and State guidelines[659]. Similar orders for testing of AVs have been issued during 2016-18 by Governors of Massachusetts, Minnesota, Delaware, Hawaii, Ohio and Idaho.

Pre-2017 Estonia had allowed the testing of AVs on the roads in the country[660]. Post-2017 Estonia permitted the usage of AVs by the public with a condition that the AVs must have a safety driver inside who must be capable of taking control of the car on a required basis. Recently, the Finnish government passed two legislations to allow AVs[661]. The country now permits AV Taxies with a safety driver who can control the vehicle remotely. Finland presented a self-driving Robobus with a human operator in May 2018. Finland intends

---

[655]. Id.

[656]. Id.

[657]. Bill Covington, Legislating Autonomous Vehicles in Washington (2018), https://wstc.wa.gov/wp-content/uploads/2019/11/2018-0717-BP7-UWFullReportAVLawScan.pdf (last visited December 30, 2020).

[658]. Executive Order 2015-09 [online] available at https://azgovernor.gov/file/2660/download?token=nLkPLRi1(last visited December 30, 2020).

[659]. Arizona says human-free driving is a-okay https://www.theverge.com/2018/3/2/17071284/arizona-self-driving-car-governor-executive-order (last visited December 30, 2020). The article explains that as per the executive order, fully driverless cars without anyone behind the wheel are allowed to operate on public roads. The only caveat is that the vehicles follow all existing traffic laws and rules for cars and drivers.

[660]. Self-Driving Min Buses set to roam in the streets of Tallin in 2017 [online] available at https://e-estonia.com/self-driving-minibuses-set-to-roam-the-streets-of-tallinn-in-2017/ (last visited March 21, 2021).

[661]. 2019 Autonomous Vehicle Readiness Index Supra note 277, at 71.

to launch its first fleet of automated buses in 2022[662]. As of date, France plans to permit[663] Level 4 driverless AVs on its roads. Presently French laws only permit certain companies to test driverless AVs and this restriction is basically to ensure minimum risk to the public.

4.3   Automated Lane Keeping System (ALKS) Regulation

Around 50 countries, including Japan, South Korea, and 42 EU nations, have agreed on single legislation for AVs (ALKS), notably the kind of AV that may fully replace driving functions, under the auspices of the United Nations Economic Commission for Europe (UNECE)[664]. As per UNECE, this law is going to be the first legally binding worldwide standard for Level 3 AVs and above (driverless) thereby helping the world achieve safe and sustainable mobility for all. The EC has mentioned that these rules would take effect at a later date in the EU. Even though the United States of America is not part of the forum, it has no option but to follow the regulations in case it wishes to sell its AVs in Japan[665]. ALKS can be activated only when the safety driver is behind steering with a seatbelt on and only on roads which has a divider and where pedestrians and cyclists are prohibited.

4.4   Advanced Driver Assistance Systems (ADAS)

Advanced driver-assistance systems (ADAS) as the name suggests use advanced technologies in vehicles to help the driver. These could include many safety features. ADAS could also use sensors in a vehicle like radars and cameras and then analyse the collected information and provide it to the driver if there is a safety driver in the AV else take automated actions as per the inbuilt software

---

[662] New Autonomous Shuttle Bus unveiled by Finland's 4 OY and Moove GMBH [online] available at https://www.autofutures.tv/2021/12/02/new-autonomous-shuttle-bus-unveiled-by-finlands-sensible-4-oy-and-moove-gmbh/ (last visited March 21, 2021).

[663] France pushes for 'highly automated' vehicles by 2022 Automotive News Europe, https://europe.autonews.com/article/20180808/ANE/180809840/France-pushes-for-highly-automated-vehicles-by-2022 (last visited March 2, 2021). The government has made 40 million euros ($46 million) available to help subsidize new projects.

[664] Countries agree regulations for automated driving Tech Xplore - Technology and Engineering news, https://techxplore.com/news/2020-06-countries-automated.html (last visited April 2, 2021).

[665] Id.

program[666]. ADAS also functions as an active safety system. It can efficiently and actively control an AV's braking or steering. That's one of the reasons why the word assistance is included in ADAS. All the above features help ADAS save human lives. Studies have shown that collision warning systems have reduced crash rates by 27% and if automatic braking was included this number is doubled. Similarly, rear-view cameras help reduce crashes while reversing by 17% and including automatic braking while reversing increases this number to 78%[667].

4.5 Autonomous Vehicle Readiness Index (AVRI)

Over 30 countries utilise the Autonomous Vehicles Readiness Index (AVRI), a global methodology, to assess the readiness of AVs. AVRI is a comprehensive index which measures 28 individual scores from a variety of sources to translate them into a single score. In most countries, AVRI is generally used by public sector organizations which are responsible for transport and infrastructure. AVRI is also used by other public and private sector entities that engage in or rely on road transportation[668].

4.6 Fatal AV Crashes

The first U.S. Autopilot fatal accident involving Joshua Brown, 40, of Canton, Ohio, occurred on 07 May 2016 at Williston in Florida. According to NHTSA, the tractor-trailer turned left in front of the Tesla car, causing the crash. Evidence shows that the driver failed to use the brakes before the collision occurred at a crossroads on an uncontrolled access road[669]. The National Transportation Safety Board (NTSB) released its report in September 2017, concluding that the truck driver's refusal to give way to the automobile and the car driver's inattention as a result of an overreliance on vehicle automation were the probable causes of the collision in Williston, Florida[670]. The second US Tesla autopilot fatal crash occurred on 23 March 2018 in Mountain

---

[666.] 2020 Mobility Insider October 02, What Is ADAS? Aptiv, https://www.aptiv.com/en/insights/article/what-is-adas (last visited April 28, 2021).

[667.] Id.

[668.] 2020 Autonomous Vehicles Readiness Index, Supra note 403, at 109.

[669.] Tesla Autopilot, WIKIPEDIA (2021), https://en.wikipedia.org/wiki/Tesla_Autopilot#Handan,_China_(January_20,_2016) (last visited January 2, 2021).

[670.] Id.

View, California. The collision happened early in the day at the highway 85 carpool lane exit when the car hit a concrete barrier and as a result, two more following vehicles crashed behind it[671]. Thereafter the car caught fire. There is no publicly available news on any compensation paid by Tesla to Mr. Huang's family nor is there any news on any product liability case against Tesla.

The third US self-driving vehicle crash was by an Uber vehicle in Tempe, Arizona in March 2018, which was a famous case that hit the headlines. The vehicle in this case was a Volvo XC90 car and preliminary investigations revealed that the speed of the vehicle required an emergency braking system to work. Uber had disabled the factory settings for emergency braking to lessen the possibility of unpredictable vehicle behaviour. However, National Transportation Safety Board (NTSB) did not apportion any blame to Uber in their preliminary report[672]. After conducting an investigation, the NTSB assigned culpability to the State of Arizona, Uber, the safety driver, the victim (who had tested positive for methamphetamine and marijuana in addition to crossing the road in an unauthorised location), and the safety driver. The NTSB report mentioned the failure of the federal government to properly regulate the Autonomous Vehicles industry. Uber was vindicated of any criminal wrongdoing by local authorities. Some of the other fatal AV crashes are mentioned in subsequent paragraphs.

On 20 January 2016, a Tesla car driver Gao Yaning was killed in Handan, China after his vehicle collided with a stationary truck. The Tesla car was plying, behind another car which suddenly changed to the right lane for avoiding a truck parked on the left shoulder which the Tesla car in autopilot mode did not detect. In the lawsuit filed by the driver's family, Tesla responded by saying the car was damaged heavily and hence telemetry could not be retrieved to understand if autopilot was engaged or not[673]. The lawsuit was stalled in 2018 due to issues in retrieving telemetry data. Later the deceased driver's family managed to retrieve the information and they sent that to Tesla, post which

---

[671.] Tesla says Autopilot was active during fatal crash in Mountain View [online] available at https://arstechnica.com/cars/2018/03/tesla-says-autopilot-was-active-during-fatal-crash-in-mountain-view/ (last visited January 2, 2021).

[672.] Andrew J. Hawkins, Uber self-driving car saw pedestrian but didn't brake before fatal crash, feds say The Verge (2018), https://www.theverge.com/2018/5/24/17388696/uber-self-driving-crash-ntsb-report (last visited December 30, 2020).

[673.] Tesla admits autopilot feature led to fatal China crash in 2016 [online] available at https://www.yicaiglobal.com/news/tesla-admits-autopilot-feature-led-to-fatal-china-crash-in-2016 (last visited March 24, 2021).

Tesla acknowledged that two minutes before the collision, the autopilot was activated. After this, Tesla took "Autopilot" off of its Chinese website[674].

On 01 March 2019, in Florida, a Tesla Model 3 hit a semi-trailer truck which was turning left. The impact was so huge that the greenhouse of the Model 3 was sheered, and the Tesla driver Jeremy Banner died[675]. Investigations revealed that the Tesla driver had engaged the autopilot just ten seconds before the crash. Also, the telemetry results did not show the driver's hands were on the wheel for eight seconds before the collision. On 29 May 2020 a Tesla car in Arendal, Norway, hit a truck driver, who was positioned close to his semi-trailer in the street (which was partially off the road). Charges of negligent homicide were brought against the Tesla driver and he was finally sentenced to three months' imprisonment in December 2020[676]. On 29 April 2018, a Tesla car driving on Autopilot mode struck and killed a forty-four-year-old pedestrian in Kanagawa, Japan. The driver of the Tesla car was asleep at the time of the accident. A lawsuit has been filed in a federal court in Northern California in April 2020[677]. The lawsuit[678] has apportioned the cause of the accident to a defect in Tesla's autopilot system including little enforcement of careless driving.

On 05 November 22, a Tesla Model Y car (which falls in the Autonomous Car category), with a driver on the steering wheels, slowed down to park and then suddenly picked up speed and killed two persons, one of which was a school girl. The driver has claimed that he tried to apply brakes and they malfunctioned and apparently, the vehicle picked up speed. Tesla on the other hand relying on the video footage claimed that no brake light was seen. However, they have committed to the Chinese government that they will support the investigation. This crash happened in the southern province of Guangdong in China[679].

---

674. Id.

675. Tesla's Autopilot was engaged when Model 3 crashed into truck [online] available at https://www.theverge.com/2019/5/16/18627766/tesla-autopilot-fatal-crash-delray-florida-ntsb-model-3 (last visited January 2, 2021).

676. Norway Today [online] available at https://norwaytoday.info/news/tesla-driver-charged-with-negligent-homicide-of-truck-driver-he-hit-while-using-auto-steering-on-e18/ (last visited March 21, 2021)

677. Tesla Autopilot Technology Killed a Man in Japan, According to This Lawsuit [online] available at https://www.motorbiscuit.com/tesla-autopilot-technology-killed-a-man-in-japan-according-to-this-lawsuit/ (last visited July 7, 2021).

678. Id. The lawsuit, filed in Tesla's home state of California, alleges that Tesla willfully ignored the NTSB's recommendations for safety regulators, and is seeking damages on behalf of the victim's family.

679. Fatal Tesla car crash caught on CCTV in China available at https://www.hindustantimes.com/world-news/tesla-car-accident-in-china-china-tesla-accident-news-video-fatal-tesla-car-crash-caught-on-cctv-in-

4.7 Non-Fatal AV Crashes

The first non-fatal crash in the USA was one was on 22 January 2018 in Culver City, California wherein a Tesla Model S hit a stationary fire truck after the automobile in front of it switched lanes and the Tesla's sensor/autopilot failed to detect the stationary truck in time[680]. The NTSB in their report submitted on August 2019 concluded the driver's fault and based it on the driver's negligence and excessive reliance on the modern driver assistance system of the car. The second non-fatal crash occurred in South Jordan, Utah on 11 May 2018 again with a Tesla Model S which hit a stationary truck. In this case, the driver of the Tesla was injured in her foot, and she admitted to looking at her phone just before the crash[681]. NHTSA reports indicate that "the driver did not touch[682] the steering wheels at least for eighty seconds preceding the crash" and the driver was solely blamed for the accident.

The third non-fatal crash occurred on the evening of August 10, 2019, when a Tesla Model 3 that was travelling in the left lane of the Moscow Ring Road in Russia collided with a tow truck that was parked there with a corner sticking out into the lane and then caught fire. The driver stated that while his hands were on the wheel, he was not paying attention at the time of the collision and that the car was moving at the posted speed limit of 100 kilometres per hour (62 miles per hour)[683]. The fourth non-fatal crash occurred on 01 June 2020 in Taiwan when a tesla went and hit an overturned cargo truck. While this was happening the traffic, cameras captured the entire accident. The driver informed emergency personnel that the vehicle was in Autopilot mode and that he was unharmed. According to reports, the driver admitted to police that he saw the vehicle and manually applied the

---

china-schoolgirl-among-2-killed-101668414061237.html (last visited November 14, 2022).

[680]. Tesla car was on Autopilot when it hit a Culver City firetruck, NTSB finds [online] available at https://www.latimes.com/business/story/2019-09-03/tesla-was-on-autopilot-when-it-hit-culver-city-fire-truck-ntsb-finds (last visited January 2, 2021).

[681]. Driver in South Jordan auto-pilot crash sues Tesla [online] available at https://www.deseret.com/2018/9/5/20652799/driver-in-south-jordan-auto-pilot-crash-sues-tesla (last visited January 2, 2021).

[682]. Id.

[683]. Tesla Explodes After Crashing into Tow Truck While on Autopilot [online] available at https://www.newsweek.com/video-tesla-explodes-tow-truck-1453745 (last visited March 21, 2021).

brakes too late to prevent the collision[684]. This statement of the driver[685] is also corroborated by the recorded video where one could see white smoke emerging from the tyres in a puff.

4.8 Conclusion

There are many countries in the world which have no laws for AVs and some of these prominent ones are Slovakia, Philippines, Bulgaria, Turkey, Romania, Ukraine, Indonesia, Malaysia, Vietnam, Australia, and New Zealand.

Vehicle Platooning laws are mainly found in various State laws in the USA. Some of the salient features in these laws are the requirement of mandatory permission before platooning vehicles, removal of minimum distance or "too close" rule, use of vehicle-to-vehicle communication technology and use of an electronically coordinated braking system to ensure the appropriate distance between platooned vehicles. Apart from various states in the USA, Norway is the only other country in the world, a non-US country that has permitted testing of AVs on its roads and simultaneously started working on truck platooning from 01 January 2018.

Various States in the USA have approved the testing of AVs only by certain qualified organisations. Some states in the USA require prior permission before testing. This precedence has been followed in other countries across the world like Austria, Hungary, Italy, Monaco, Netherlands, China and Singapore. The testing areas, speed limits etc are also specified in the government orders and approval for testing.

Analysis of the definition of AVs across the world reveals that most of the definitions are almost similar and they talk about a vehicle equipped with an automated driving system that is intended to operate without a driver. The SAE levels are also often referenced in the definition of an AV and a vehicle must be of SAE Level Four or Five to qualify as a fully autonomous vehicle. The Russian regulations also mandate that every autonomous vehicle must be marked with a big, black A, mentioned on it to distinguish it as an AV. The

---

[684.] Tesla In Taiwan Crashes Directly into Overturned Truck, Ignores Pedestrian, With Autopilot On [online] available at https://www.forbes.com/sites/bradtempleton/2020/06/02/tesla-in-taiwan-crashes-directly-into-overturned-truck-ignores-pedestrian-with-autopilot-on/?sh=46729fd658e5 (last visited July 7, 2021).

[685.] Id.

most prominent difference between the US regulations and that of Germany is in the definition of AVs. In the USA Level 5 AVs can be driverless. But as per StVG in Germany, there is a need to have a safety driver in all AVs and they must have the capability to take over control on a required basis.

Most jurisdictions across the world where AVs are permitted to ply or tested require the AV to be insured for at least One Million USD to Five Million USD. Nebraska mandates that manufacturers of automated driving systems are required to carry $1 million in liability insurance and $5 million in general liability insurance. Bulgaria has a mandatory requirement for third-party liability insurance. In Italy, civil liability cover for AV must be equal to four times the cover for a non-AV vehicle. Wherever law does not specifically call out AV insurance there the general automobile insurance guidance is followed.

Proper infrastructure support has, is and will always be a necessity for AVs to ply properly and safely and most countries across the world which are in favour of unmanned AVs are working towards the same. Similarly, there are quite a few states in the USA and many countries across the world, that mandate the presence of a driver in an AV, and there are many which do not mandate it and permit remote drivers. Some of the countries which mandate the physical presence of drivers are Norway, Portugal, Russia, China, and Italy.

When it comes to product liability most states in the USA which permit AVs to ply have well-defined product liability laws in place. Most of these rules shield car technicians and repair shops from product responsibility when they work on automated vehicles per manufacturer guidelines. Immunity is also generally provided from civil liability to the manufacturer of automated technology for damages caused due to third-party alterations made to a motor vehicle/automated motor vehicle. In most jurisdictions, product liability also applies to software that's part of a physical product like an AV. Wherever specific product liability rules are not specified the general liability rules would apply to the manufacturers. Law relating to recalling defective vehicles exists in many jurisdictions and China. The guarantee liabilities come in three buckets, and they are guaranteed liability for repair, replacement, and return.

The importance of data generated by AVs is well known and most jurisdictions have laws to protect the same. The requirement of recording devices and processes for handling and storage of data including deletion is spelt out in many laws. Some of the notable ones are Alabama, Utah, Germany, and Russia. As per Russian regulations, every AV must mandatorily have a data recording system and the driver must also be able to switch off and on the equipment for traffic recording and actions of the driver. The Automated Lane Keeping System or ALKS of the UNECE has a requirement to have a Blackbox in all AVs which can record every minute detailed recording of the AV's journey and the way the AV was handled during the journey or an accident. In Indonesia, the Data Protection Regulation prescribes that any offshore transfer of personal data can be made only after due approval.

The Society of Automotive Engineers in the USA popularly known as SAE has designed a zero to five rating system to categorise different levels of automation. Greater automation means a higher level and lower automation means a lower level. Level 1 and 2 is allowed in most countries. Level 3-5 is restricted and allowed only in a few countries across the world. Russia, Slovenia, and Ukraine, as of date do not allow Level 3 to Level 5 on their public roads. China has standards akin to SAE levels. Quite a few states in the USA allow upto Level 5 for testing and even use.

Accidents caused by AVs are complex to dissect when it comes to liability and different jurisdictions across the world have afforded different treatment to this complex aspect of AV accident liability. Colorado, Nevada, Austria, Bulgaria, Germany, Monaco, Poland, Portugal, Switzerland, Turkey, Singapore and Ukraine attribute liability based on existing laws as they do not have separate laws for AVs. As per Tennessee laws, if an AV is operating in full automation mode, then that AV will be considered a driver for any offence. In Austria, Russia and Slovenia the driver/operator/owner can be made liable for any AV accidents. In PRC an Entity carrying out ICV testing must have sufficient financial capability to pay up any kind of compensation due to personal or property damages caused by ICV testing.

The Self-Drive Act of the USA was enacted to mainly adopt federal safety standards in the arena of AVs. Post that a bill called AV START Act was moved to set up a Highly Automated Advisory Council to assess self-driving car-related challenges and discusses federal regulation on AV testing. But this bill was never passed. Around 50 plus countries including Japan, South Korea and around 42 EU countries have agreed on a common regulation for AVs (ALKS) especially the kind of AV which can completely take over driving functions. As per UNECE, this law is going to be the first legally binding worldwide standard for Level 3 AVs and above (driverless) thereby helping the world achieve safe and sustainable mobility for all. Advanced driver-assistance systems, sometimes known as ADAS, are computer programmes that support drivers. By using sensors in a vehicle like radars and cameras and then analysing the collected information and providing it to the driver if there is a safety driver in the AV else take automated actions as per the inbuilt software program. It can be safely concluded that for any successful AV operation ALKS, AVRI and ADAS are essential requirements, legally and infrastructurally.

* * *

CHAPTER 5

CHANGES REQUIRED IN LAWS OF INDIA TO PERMIT AUTONOMOUS VEHICLES

5.1 Background of Autonomous Vehicles in India

It is a well-known fact that there are many automobile manufacturers globally who are working on autonomous vehicles or AVs. As per SAE standards, Level 5 AV is a completely driverless or self-driving vehicle. When we say completely driverless and self-driving, what it implies is that these AVs can sense the surrounding environment and can easily navigate the vehicle without the intervention of a human driver. It is a well-known fact that India is completely lagging in the adoption or initiation of steps to adopt AV technology. India has not even commenced the baby steps that many Asian nations have commenced.

India is completely aware of the following factors/data concerning AVs, some of which have been described in previous chapters. At least 90% of automobile crashes are caused by human error[686]. Self-driving AVs will help reduce human error and thereby contribute to reducing the accident count. Driverless AVs can help reduce fuel/power consumption as these vehicles can talk to each other using vehicle-to-vehicle (V2V) communication and thereby

---

[686.] What percentage of car accidents are caused by human error? [online] available at https://www.bttlaw.com/what-percentage-of-car-accidents-are-caused-by-human-error/ (last visited July 7, 2022)

using their intelligence on gathered data to adopt a route with less traffic. AVs are very useful in taking care of the needs of specially-abled persons. AVs can help such persons travel anywhere without depending on other humans.

AVs can also help curb the menace due to the issuance of fraud/bogus driving licenses, especially in countries like India. In tune with the UN SDG, most countries are now slowly moving to EVs to control pollution. The present trend is to blend AV into EV. AVs can also help provide safe travel for women and ensure their safety, which is very much essential in countries where women find it unsafe to travel alone. Driverless AVs can also be used as delivery vans or for pickup. This will help conserve manpower and time.

Despite the above, a few years back one of the Road Transport Ministers in India proclaimed that he is not in favour of introducing driverless cars in India for the main reason that it will propagate loss of jobs for around 2.2 million commercial drivers. He mentioned that his focus would be to train drivers[687] and equip them with adequate skills, which will not only help create adequate jobs but also improve the public transport system in India. Just a year before this announcement, Mahindra and Mahindra (M&M) automobiles of India exhibited their indigenously manufactured autonomous tractor[688]. M&M had mentioned that they have plans to make these tractors going forward. It is quite evident that to make a self-driving AV that can run in India the automobile OEM will need millions of kilometres of road data and also develop an algorithm that will work in Indian conditions. As of date apart from the above announcement by M&M no other automobile OEM in India has made any announcement about AV for India.

In a poll conducted in India, a good percentage of Indians expressed that they are in favour of self-driving AVs[689]. Many also expressed their desire to lease or buy an AV considering the huge traffic congestion in the country. Mumbai is soon going to have driverless Metro which is already there in many countries. The expectation is that post-implementation of driverless metros,

---

[687] Group Discussion Ideas. 2021. *Should Driverless cars be allowed in India? ~ Group Discussion Ideas.* [online] Available at: <https://www.groupdiscussionideas.com/should-driverless-cars-be-allowed-in-india/> [Last visited August 17, 2021].

[688] Moneycontrol. 2021. *Autonomous Driving Technology | Can Self-Driving Cars Run on Indian Roads | Moneycontrol.* [online] Available at: <https://www.moneycontrol.com/news/driverless-cars/> [Last visited September 17, 2021].

[689] Indians most keen on driverless cars [online] available at https://www.autocarindia.com/car-news/survey-indians-most-keen-on-driverless-cars-409013 (last visited September 17, 2022)

self-driving AVs can get implemented in areas of public transport and logistics. Considering the topography of India and its population density, it is a no-brainer that to implement an AV ecosystem within India, there will be a great amount of investment required. India has been working on EVs for quite some time to tackle pollution problems. There are many states in India which ply EV buses as public transport[690]. Since India has been the innovation hub for Silicon Valley, there is no doubt that with the background of EV innovation and IT innovation, India can be a hub for AV innovation and export as most nations are looking to integrate IT capabilities and EVs into AV.

In 2020 KPMG released a report depiction the readiness index for AVs also known as Autonomous Vehicle Readiness Index[691] (AVRI). This report analysed around thirty countries on four aspects namely policy and laws, technology and innovation, infrastructure, and customer acceptance. India's position in this report was very bad. It stood last in customer acceptance and AV regulations and just above Brazil in infrastructure availability. As far as road quality goes, India was in the bottom ten lists along with Mexico and the bottom five for innovation. The report also mentioned that the current focus in India is to develop the use of EVs before supporting AV development. Also, the adoption of AVs and EVs could influence the Insurance and IT sector.

Since India is lagging in the AV field it would be good to go through the various existing legal provisions in India for Motor Vehicles and also try to understand if they can be used as is for Self-Driving AV operations or need modifications/amendments.

5.2 Motor Vehicles Act 1988 and Amendment Act 2019

The Motor Vehicle Act,1988 (MV Act)[692], implemented on 01 July 1989, is an all-encompassing law on all issues connected with motor vehicles

---

[690] Future of Public Transports in India – Electric Buses [online] available at https://www.financialexpress.com/auto/electric-vehicles/future-of-public-transport-in-india-electric-buses/2421366/ [Last visited August 17, 2021].

[691] 2020. *Autonomous Vehicle Readiness Index.* [ebook] Available at: <https://assets.kpmg/content/dam/kpmg/uk/pdf/2020/07/2020-autonomous-vehicles-readiness-index.pdf> [Last visited September 17, 2021]. The KPMG report also mentioned that the initial usage of AVs and EVs could mainly be in warehouses, ports, and farms. It is said that usage of AVs in farms can reduce dependency on migrant workers which we saw as a key factor during the pandemic.

[692] Mukherjee, T., 2006. *Motor Vehicle Act, 1988 with rules and exhaustive commentary on motor accident compensation.* Allahabad: Premier Pub. Co.

and road safety. The prior acts were the Motor Vehicles Act, of 1939 and the Motor Vehicles Act, of 1914. The MV Act contains legal provisions concerning driver's licenses, conductor's licenses, vehicle registration, vehicle permits, traffic rules, vehicle insurance, driver/owner liability, offences, and penalties. The MV Act[693] has various provisions to claim compensation for any death or injury caused to a person on account of any accident. The 1994 amendment[694] to the MV Act created special provisions under Sections 66 and 67 to exempt vehicles operating on eco-friendly fuels from the requirements of the permit. This proviso also permitted owners of such vehicles to decide on the fare and goods freight charges. The whole aim of this amendment was to support the influx of vehicles working on eco-friendly fuels. As regards judicial precedence, there are a plethora of cases concerning the interpretation of various sections of the MV Act and Rules. The Motor Vehicles Act was later amended in 2019 vide Motor Vehicle (Amendment) Act 2019 which was enforced from 01 September 2019[695]. Vide the 2019 Amendment[696] penalties were also increased to bring about a deterrent factor.

Quite a few amendments were made in 2019 to bring the MV Act to par with global standards. Some of these amendments, which can also be called stepping stones for future amendments to facilitate self-driving AV operations in India are discussed in the subsequent paragraphs. The 2019 amendment permits vehicle users to carry digital soft copies[697] of their car papers and driver's license. As per the earlier rule, hard copies of these papers/licenses were required to be carried. Applications like Digi Locker and M-Parivahan have been notified by the government as authorised storage space for storing soft copies of these documents. This amendment can also help in retrieving vehicle documents from Self-Driving AVs if suitable digital systems are installed in these vehicles. That will help trace the owner easily in case of accidents, apart from using the conventional route of tracing through vehicle registration numbers.

The Amendment Act has introduced heavy fines and punishment for offences like drunken driving, driving without possessing a license and

---

693. Id.

694. 2019. Manual of Motor Vehicles Laws. Commercial Law Publishers.

695. 2019. *Motor Vehicle Amendment Act 2019.* [ebook] Government of India. Available at: <https://egazette.nic.in/WriteReadData/2019/210413.pdf> [Last visited August 17, 2021].

696. Id.

697. Id.

speeding. The amendment act is also futuristic in the sense that it prescribes these penalties/fines to be increased by 10% every three years[698]. As per the amendment, a driving license can now be renewed upto one year from the date of expiry as against the previous one-month period. But delay beyond the prescribed one-year period now mandates a test of competence before issuing the license[699]. When it comes to AVs this provision may have to be relooked as one year would be a long time for technology to have changed and the currency of the driver on the AV technology should be considered an important factor before renewing an expired license. Else the law should prescribe some kind of continuity training before agreeing to renew an expired license of an AV driver.

The amendment act has also increased the minimum compensation for death or serious injury under Section 161[700]. For death, it is now rupees five lakhs as against rupees fifty thousand earlier and for grievous hurt, it is rupees two and a half lakhs as against rupees twenty-five thousand earlier. The Amendment Act also protects "Good Samaritans", as defined in Section 134A[701] of the Amendment Act, from any liability for acts caused by their deeds done in good faith including while transporting the accident victim to a hospital. One good thing about the 2019 amendment is the aspect of transparency as every violation will be monitored online including uploading of accident photo clips and videos. This revision in the Amendment Act is a good move and the only aspect that needs to be specifically factored when AVs are permitted is the amendment to insurance clauses/requirements, else one would have legal compensation provisions without proper resources for implementation.

The 2019 Amendment Act has inserted a new sub-section 1A under section 138[702] to enable the State Governments to create cycle tracks, footpaths, and special lanes for non-motor traffic to protect pedestrians and non–motor traffic. Similarly, the 2019 amendment inserted a new section 194B[703] to make it mandatory for a child travelling in a motor vehicle/

---

[698]. Id.

[699]. Legalserviceindia.com. 2021. *Motor Vehicle Act, 2020 (Amended)*. [online] Available at: <http://www.legalserviceindia.com/legal/article-4188-motor-vehicle-act-2020-amended-.html> [Last visited August 18, 2021].

[700]. Motor Vehicle Amendment Act 2019, Supra Note 707, at 189.

[701]. Id.

[702]. Id.

[703]. Id.

motorcycle to wear a child restraint system, safety helmet, or both with an appropriate fine for any failure. The 2019 amendment[704] mandates that vehicles that don't satisfy certain standards will be recalled by the central government. All the above legal provisions relate to the safety of individuals and would be paramount even during self-driving AV operations. Road design and its maintenance are part of the mandatory infrastructure requirements for the safe and successful implementation of self-driving AV programs in any nation. Towards that India has taken baby steps through new Section 198A[705] inserted vide the 2019 amendment act, which helps to hold road consultants and contractors responsible for poor designing of roads, their construction and bad maintenance.

Another new feature in the 2019 amendment, which can be adopted in the amendments to be made for AV laws is the aspect of permitting a person directly apply for a driving license if he/she has received accredited school training for that vehicle class. The new Section 25A[706] in the 2019 amendment now caters to maintaining a National Register for all driving licenses issued across India and all licenses to have a unique number from this register else they will be invalid. The mandate is that the state registers must get merged into the central register. These amendments will help digitize the entire motor vehicle regulatory ecosystem and that is something which is very much essential to be in place once we have self-driving AVs plying in India.

While the 2019 amendment has prescribed age limits for licenses and also for renewal of licenses, apart from aspects like maximum passenger limit, maximum luggage weight and the number of permitted trips (max and min) for commercial vehicles, the amendment has not provided or accounted for fatigue tests, the construction of rest facilities for transport drivers, the training of transport drivers through simulated tests, and the minimum necessary safety standards for vehicles. These fatigue tests and safety standards regulatory requirements are very much essential in the space of self-driving AVs too. Hence the earlier we incorporate them into our existing laws, the easier it is to adopt them for AVs. Another important aspect of road safety and regulation for AVs is electronic monitoring of road traffic and an element of this can be found in the new Section 136A inserted vide 2019 which now casts a responsibility on

---

704. Id.
705. Id.
706. Id.

the Central Government[707] to make and implement effective regulations for computerised traffic safety enforcement. Implementation of speed cameras, CCTV cameras and speed guns will help capture violations at a greater scale and also provide proper evidence in case of any accidents including accidents by self-driving AVs as and when permitted to drive in India.

Some of the other futuristic amendments implemented through the 2019 amendment which can be looked at as supportive of the AV regime in India is the amendment to the old driving dangerously[708] definition by bringing more clarity to state that it includes dangerous acts like jumping signals when red, use of mobiles, violation of stop signs and driving in opposite direction[709]. The violation of jumping signals and opposite-direction driving can also happen with self-driving AVs either due to remote driver's faults or faulty programming. In most of the states in the USA and Europe, we have seen that government-owned Road Safety Boards or similar organisations are of paramount importance while deciding self-driving AV policies or even evaluating accidents involving AVs. As a stepping stone, the 2019 amendment through new Section 215D has established a National Road Safety Board[710]. Also, Sections 66A and 66B of the 2019 amendment, empower Central Government to consult with the States towards working and implementing a National Transportation Policy. The mandate of the National Road Safety Board is to render advice to Central and State governments on road safety, management of traffic, design of roads, maintenance of vehicles and roads, construction standards and technology for roads and motor vehicle standards. In the years to come, when India will gear up to permit AVs, the role of this Board can be expected to advise on all aspects of AVs too. Cab Aggregators are also in the future expected to use AVs and hence the need that these aggregators also must be regulated. As of now since India did not have any law for Cab Aggregators, the 2019 amendment has done a good thing by recognizing such aggregators under Section 93[711]. Finally, another new change brought about by Section 56 of the 2019 amendment, which provides for testing of vehicle fitness through automated testing centres is also futuristic and would be very useful post-AV implementation regime in India.

[707.] Id.

[708.] 2019. Manual of Motor Vehicles Laws. Commercial Law Publishers, Supra Note 706, at 189.

[709.] Motor Vehicle Amendment Act 2019, Supra Note 707, at 189.

[710.] Id.

[711.] Id.

5.3 Motor Vehicle Insurance and Case Laws

Existing legal regime for AVs in various states of the USA and Europe indicates that Insurance is a prime factor not only when permitting AVs to ply but also when permitting testing of AVs. The reason is, AVs can sometimes behave like unguided missiles and cause huge losses in terms of money and life. An insurance contract is one in which the insured party pays the insurer a sum of money (referred to as a premium) under the terms of the agreement, and the Insurance Company undertakes to reimburse the Insured in the event of any accident/specified uncertain events. Insurance has its roots in the fourteenth century wherein maritime risks were insured. Later it spread to life and fire insurance. As per the MV Act, every vehicle plying on Indian roads must be Insured. Section 140 of the MV Act[712] introduces a term called "No Fault Liability". As per this section even if the accident or injury or loss or death or disability is caused without the driver or vehicle owner's fault, the vehicle owner has to provide compensation. The no-fault liability clause mandates paying a fixed sum as compensation depending on what has been caused by the accident (death or disablement)[713]. Needless to reiterate as the name of the provision is no fault liability, aggrieved parties need not prove any wrongful act or neglect by the defendant. A landmark case on no-fault liability that is always referred to is National Insurance Company vs Deorao and others decided in 2002.[714] In this case, the court upheld the provisions of no-fault liability and said that this provision aims to compensate without any question. In the case of AVs most countries/states in the USA mandate, a prescribed sum of insurance during testing and playing to compensate for similar insurance claims. The existence of no-fault liability in the Indian Motor Vehicle Act is supportive of the futuristic insurance requirements for AVs to play in India. The only aspect is the coverage amounts may have to be upwardly revised.

One can commonly see people mixing up no-fault liability with strict liability. The compensation in cases of no-fault liability is set by legislation, whereas in the case of strict liability, it comes under the law of torts and there is no prescribed compensation amount. As per Section 140 MV Act 1988,

---

[712] 2019. Manual of Motor Vehicles Laws. Commercial Law Publishers, Supra Note 706, at 189.

[713] Linkedin.com. 2019. *No Fault Liability*. [online] Available at: <https://www.linkedin.com/pulse/fault-liability-major-prashant-rai/> [Last visited August 18, 2021].

[714] II (2004) ACC 328

no-fault liability for death[715] is Rupees fifty thousand only and for disability is Rupees twenty-five thousand only. Another mandatory requirement under the MV Act is third-party liability or third-party insurance coverage. This is also commonly known as act-only cover. This kind of insurance is called third-party insurance, as the beneficiary is a party who does not participate in the insurance agreement. A third-party insurance policy takes care of the insured persons' liability for death/disability/damage to property caused to third parties. Third-party insurance cover is fault-based, which means firstly one must prove the fault of the insured and then connect that injury with the fault of the insured. Only post that one can claim damages from the insured. Insurance companies are not very happy with the third-party insurance policy as it is like an unlimited policy and has no cap.

One of the landmark judgements defining what a "Third Party" means is the case of National Insurance Co Ltd vs Fakir Chand reported in AIR 1995 J&K 91. This judgement stipulated that while Section 145 of the MV Act mentioned that Government is included in the definition of the third party, the court found that the third party shall consist of all individuals not a party to the deal e.g., a person travelling in another vehicle, a person walking on road and passenger in a vehicle (as was the dispute in the instant case)[716]. Another landmark judgement on third-party insurance, covering the mandatory requirement of third-party insurance in place for all motor vehicles was delivered by the Hon'ble Supreme Court in the case of G Govindan vs New India Assurance Co Ltd reported in AIR 1999 SC 1398[717]. The moot question, in this case, was, if the Insurance Policy has lapsed, will the owner's liability cease, if the insured vehicle is transferred and no application or intimation for change of name of the insurer is made under the motor vehicle act. The High Courts of Delhi and Karnataka had in their judgements in different cases held that the liability ceases, while the High Court of Andhra Pradesh (all Full Bench judgments) had mentioned that the liability does not cease, and it continues. The Andhra Pradesh (AP) Judgement was in the case of Madineni Kondaiah and Ors. v. Yaseen Fatima and Ors[718], AIR (1986) A.P. 62. In the AP case, the court held that unless there is a novation agreement

---

715. 2019. Manual of Motor Vehicles Laws. Commercial Law Publishers, Supra Note 706, at 189.
716. AIR 1995 J K 91.
717. 1999 (2) SCR 476.
718. Madineni Kondaiah & Ors. v. Yaseen Fatima & Ors. on June 22, 1985, 1 (1986) ACC 501

between insuring company, transferor and transferee, the transferee cannot secure any personal benefit under the policy. The transferee has no remedy against the insurance company. The court also went on to discuss the statutory liability of the original owner. As per the court, it is not right to conclude that as soon as title transfer happens all statutory obligations are passed on to the new owner. As per the court[719] under Section 31 of the 1939 MV Act (akin to Section 50 of the 1988 MV Act) within fourteen days of the sale, the transferor must report the sale/transfer to the transport registering authority under whose jurisdiction the transfer has been annotated in the books and post completion of all formalities once the transport authority passes relevant orders the transfer including liability shall be completed. No action by registering authority shall mean deemed completion of transfer formalities. After completion of all these processes the liability transfers to the new owner. The court has also mentioned that the request must be accepted by the insurer for transfer.

In the case of G Govindan vs New India Assurance Co Ltd finally, the court opted to accept the judgement rendered by the Andhra Pradesh High Court in the Kondaiah case as against the ratio of Delhi and Karnataka High Courts. As per the Hon'ble Supreme Court,[720] Kondaiah's case is relevant as it protects the interest of the third party, which the court believed must have been the legislative intent. The court also clarified that in this situation, the transferee whose transferor did not follow the policy transfer procedure will not be considered a third party. What the court meant was if any damage or loss has been caused to any third party other than the transferee, then that will be covered by the third-party insurance even if the transfer of insurance has not happened. However, any delay by the transferor in applying for transfer shall not cover the transferee for any loss even though the transferee is also technically a third party to the insurance agreement between the Insurance company and the transferor.

Insurance Companies are bound to settle claims as ordered by courts of law. However, Insurance companies have innumerable defences available, to avoid paying up insurance money as mentioned in Section 149(2) of the MV Act[721]. Some of them are the usage of the vehicle for hire/reward without an

---

[719.] Ibid.
[720.] 1999 (2) SCR 476.
[721.] 2019. Manual of Motor Vehicles Laws. Commercial Law Publishers, Supra Note 706, at 189.

appropriate permit, usage for racing and speed test, usage of transport vehicle beyond limits prescribed in the permit, vehicle driven by a driver without a valid driving license or when the driver was disqualified to hold such a license and lastly when the insurance policy is void as the same was obtained by not disclosing material facts. Insurance companies cannot avoid liability for any reasons other than the above. Even in these cases, there are rulings by the Hon'ble Supreme Court to state that the Insurance company must pay up and then claim the money from the Insured owner. The onus of proof for an exemption claim is on the Insurance Companies.

Insurance settlement between insurers and the insured is covered under Section 152 of the MV Act[722]. As per the said Section 152,

"No settlement can be made by the Insuring company for any liability with a third party who is not a party to the settlement and any agreement made by insuring company with insured after they have been declared to be insolvent or insolvency proceedings have commenced, to defeat the interests of liability to third party shall be invalid and original rights will continue"[723] (Section 152 MV Act).

In Sohan Lal Passi's v. P. Sesh Reddy reported vide AIR 1996 SC 2627, the Hon'ble Supreme Court of India has held that the Insurance Company must prove that the owner knew that he had a fake license. Else there is no defence available to the Insuring company[724]. In another recent case of National Insurance Co. Ltd. vs Swaran Singh decided on 30 July 1997 and reported in (1997) ACC 377, the Hon'ble Supreme Court has reiterated that proving that the driver was not holding a driving license will not absolve the insurance company unless they can prove that the breach was in the know-how of the owner.

Section 147[725] of the MV Act enumerates the nature and coverage of the Insurer's liability. It is mandatory to protect yourself from liability for death, physical harm, or property damage to other parties. The words "Third Party" includes owner, authorised representative travelling in a vehicle and any passenger using a public vehicle. Contractual liability if any is also required to be covered under Section 147. The liability continues as long as the policy is in

---

722. Id.
723. Id.
724. 1996 SCC (5) 21
725. 2019. Manual of Motor Vehicles Laws. Commercial Law Publishers, Supra Note 706, at 189.

existence and for renewals, the liability commences from the renewal date. The onus of proof rests with the automobile owner to establish that the insurance policy exists. Similarly, if the policy is not renewed and has expired the insurer will not be liable. Once a certificate of insurance is issued then the insurer cannot escape liability. If the insurer has been defrauded by the insured, then the insurer must file a separate suit to recover the amount from the insured. In the case of Oriental Insurance Co v. Inderjit Kaur reported vide AIR 1988 SC 588, the Hon'ble Supreme Court held that[726] if an insurance company has provided coverage for a bus without being paid for it, then the insurer is bound to honour the policy. The insurer cannot have the defence that he had the right to cancel the contract. In the case of K Gopal Krishnan vs Sankara Narayanan, the Madras High Court has held[727] the lack of a requirement for a scooter owner to acquire third-party risk insurance to ensure that there is adequate coverage for any claims from the pillion rider(s). Similarly, the court also held that a private car cannot carry goods for hire and if so done and injury is caused then the insurance company shall not be liable[728].

As per Section 157(1)[729] of the MV Act after the vehicle has been transferred by the owner to another individual, the certificate and insurance policy along with its rights and liabilities is also deemed transferred to the purchaser. As per Section 157 (2), the new owner must apply within fourteen days to the insuring company to effect required changes in their records concerning the change of details of the insured. Even if there is a delay or transfer has not occurred, still, Insurance Company can be made liable. Lapsed insurance policy will not offer any coverage. In the case of Sudhakaran vs AK Francis reported vide AIR 1997 Ker 26, it was held[730] that if the owner has not transferred the insurance to the new purchaser, then the insurance provider is not required to provide compensation to the purchaser or the new owner.

All insurance claims are adjudicated in India by the Motor Accident Claims Tribunal which is established at different locations pan India. Insolvency of the insured owner is no excuse, and the claim must be paid by the insuring

---

[726.] 1998 SCC (1) 371.

[727.] AIR 1968 Mad 436.

[728.] Legalserviceindia.com. 1999. *The Motor Vehicles Act, 1988. An Analysis.* [online] Available at: <https://www.legalserviceindia.com/legal/article-2663-the-motor-vehicles-act-1988-an-analysis.html> [Last visited September 1, 2021].

[729.] 2019. Manual of Motor Vehicles Laws. Commercial Law Publishers, Supra Note 706, at 189.

[730.] AIR 1997 Ker 26.

company or by the vehicle owner if there is no insurance. The above case laws and various insurance provisions indicate that Indian courts have handled all kinds of cases when it comes to Motor Vehicle Insurance and compensation and India has a great amount of precedence case laws which will come in handy when handling Insurance claim cases during AV regime, as and when permitted in India. However, when we look at the motor vehicle insurance laws in other jurisdictions, India would be required to make a minimum of following amends to the motor vehicle insurance provisions that increase no-fault liability compensation amounts for death and injury, introduce a regime of insurance for testing of AVs and introduce a different class of insurance for AVs based on their SAE levels with max insurance for SAE "5" vehicles and lowest for SAE "0" vehicles. The insurance rates should act as a deterrent but at the same time must not prohibit the introduction of AVs.

5.4   Other Allied Laws

For smooth implementation of self-driving AVs in India, apart from amending the existing MV act and rules, one will also have to look at various allied enactments which directly or indirectly affect the smooth transition of India's motor world from Non-AV to AV regime. Some of the well-known statutes that we will do a deep dive into are the Fatal Accidents Act, National Highway Act, NHAI Act, Public Liability Insurance Act and many more.

The Fatal Accidents Act of 1855 was enacted in India by the British after they enacted a similar act in 1845 in Britain known as English Fatal Accidents Act 1845[731]. This Act helps personal representatives of the deceased to file suit for damages in a court in cases where the person who caused the death of the deceased is also dead. The scope of representatives is larger in the British Act and includes brother, sister, uncle, and aunt, which is not the case in Indian law. To correct these gaps a revised bill[732] named Wrongful Death Bill 1985 superseding the Fatal Accidents Act 1855 was prepared and presented to the Parliament, which is yet to be passed. Analysis of this Act concerning the AV regime indicates that the provisions of this act can be used by relatives of the deceased to claim compensation in AV accidents, in cases where the person

---

731. 2019. Manual of Motor Vehicles Laws. Commercial Law Publishers, Supra Note 706, at 189.

732. ResponsiveWebInc, THE FATAL ACCIDENTS ACT 1855 BARE ACTS LIVE (1985), http://www.bareactslive.com/LCR/LC111.HTM (last visited September 1, 2021).

designated as an AV driver would have also died in an accident. But this Act may not be applicable if the future Indian law designated the AV itself as the driver and no human tagged as a driver. The National Highways Act and Rules have provisions which can be used by the government to acquire land for creating AV corridors in highways to enable truck platooning. Presently most trucks ply on different lanes on highways and thereby creating bottlenecks in many places. Truck Platooning is very much required in India to bring about disciplined highway driving by truck drivers and this act will help the government achieve that end as and when AVs are introduced in India. According to Section 16 of the National Highways Authority Act of India or NHAI Act[733], the NHAI is in charge of creating, managing, and maintaining all national highways as well as any additional roadways that the Indian government has given it ownership of or trusts over. Should we have special highways created for AVs or Truck Platooning the same would also likely be under the jurisdiction of NHAI for operational and infrastructure support. Minor tweaks to the existing National Highway and NHAI Act and Rules may be required to incorporate the definition of AVs and Truck Platooning.

FASTtag is a common way for collecting tolls on highways and as of now, it is done through tags affixed on the cars and in some cases paid by cash. The legal support for FASTtag is from the National Highways Fee (Determination of Rates and Collection) Rules 2008 (as amended vide GSR 920 (E) dated 25 September 2018 w.e.f 25 September 2018. These rules permit the Government to levy a fee for any section of the highway and tunnel built by a publicly-funded project as defined in Rule 2(n) of the said Rules[734]. In the case of a privately funded project, the concessionaire is allowed to collect the toll fee[735]. The 2018 amendment also made FASTtag mandatory and imposed a double penalty if a non-FASTtag vehicle entered a FASTtag lane. Once we permit AVs or self-driving vehicles to ply, then these rules will have to be further amended to cater for many aspects of the usage of fast tag like making mandatory FASTtag rules for AVs, specifying penalty and process to charge a penalty if AVs without FASTtag enter FASTtag lane or any lane if all lanes are FASTtag lanes, notification of FASTtag exemption rules for

---

[733.] 2019. Manual of Motor Vehicles Laws. Commercial Law Publishers, Supra Note 706, at 189.

[734.] Id.

[735.] NH user fee (toll) | ministry of road ... - morth.nic.in, (2019), https://morth.nic.in/toll?page=2 (last visited September 1, 2021).

AVs, if being considered by Government, connecting bank account / digital payment account of AV owners to FASTtag cards of AV to avoid any payment breach issues and process to withhold/tow AVs in case of issue in payment of FASTtag to avoid traffic jam due lane blocking.

Another statute that will be of use when AVs are implemented in India is the Control of National Highways (Land and Traffic) Act 2002 and the Highway Administration Rules 2004. This Act and the Rules under it have provisions to remove unauthorised occupation on highways and the process for the same. As per Chapter V of the said Act,[736] the Highway Administration is empowered to regulate and control the traffic plying on highways. Anytime when they feel that any portion of a highway is congested or unsafe, they are empowered to block or regulate that portion of the highway. Once we have AVs (self-driving) in India this Act will have to be amended to make special provisions concerning AVs. It is a well-known fact that AVs are programmed based on pre-existing maps, which may not contain, temporary diversions making their plying unsafe, especially if short-term diversions are created on highways say for example to solve a congestion problem due to an accident. This Act, therefore, must empower the Highway Administration to specifically come out with a different set of guidelines for driver-driven and driverless vehicles after assessing safety requirements. Rule 21 of the said Highway Administration Rules[737] also mentioned the process of handing over a vehicle or animal taken into possession by Highway Authorities. Once we have driverless cars in India the said Rule 21 will have to be amended to prescribe the process to seize these vehicles for any violation including demobilizing them, else they could be remotely controlled and can create havoc in the area where they are seized and parked especially if the owners are anti-nationals/ terrorists.

Along with the introduction of AVs globally, there is also a thrust for EVs and AVs built on EV platform is becoming more common. Most countries are aiming to promote EVs on account of environmental protection aspects to support sustainable development goals and hence offering a lot of discounts and subsidies for EVs including on their batteries which are the major component of any EV in terms of pricing. When you look at the laws in

---

[736.] 2019. Manual of Motor Vehicles Laws. Commercial Law Publishers, Supra Note 706, at 189.
[737.] Id.

India the Central Road Infrastructure Fund Act[738] and its Rules give statutory powers to levy cess/excise/customs duty on common vehicle fuels like petrol/high-speed diesel. At some stage when there is a saturation of EVs and normal petrol/diesel vehicles are off-road, the government might think of levying duties/cess on the lithium-ion batteries which are used in AVs. At that stage, this Act must be amended especially Schedule 1. Similarly, Rule 6 of these Rules[739] which talks of criteria for the selection of schemes will have to be amended to include AVs as and when we decide to permit AVs to ply in India.

Truck Platooning is unheard of in India right now, but with the arrival of AVs there is a fair possibility of Truck Platooning being introduced in India to decongest highway lanes and once that's done, we will have AV trucks also carrying hazardous substances. At that stage, the Public Liability Insurance Act and Rules and Public Liability Insurance Rules 1991, which currently assist, right away, accident victims for mishaps occurred when handling hazardous substances and for issues related to or incidental to that[740] will have to be modified to include liability elements of various actors who could be involved in an AI accident like manufacturer, owner, software coder, insurance company, municipal corporations/highway authorities for bad roads and improper/weak/nonvisible markings. Similarly, to provide better working conditions to the AV drivers[741] the Motor Transport Workers Act must be amended to consider AV drivers as Motor Transport Workers.

The Road Transport Corporations Act, of 1950 has been provided for the establishment of and control over Road Transport Corporations[742]. Presently the act defines road transport service as a business that provides the ability to transport people, products, or both, via the road for payment or reward. A tramcar, a trolley vehicle, and a trailer are all considered to be mechanically propelled vehicles that can be utilised for road transportation under this definition of a vehicle[743]. A basic analysis of the said act will infer that technically an AV when implemented into road transport service will also fall into the category of a vehicle as all AVs are mechanically propelled vehicles. Hence, the Road Transport Corporations Act will have to be amended in case

---

[738]. Id.

[739]. Id.

[740]. Id.

[741]. Id.

[742]. Id.

[743]. Id.

self-driving AVs are going to be inducted into Road Transport. The Act must clarify the definition of AV, which all AVs are permitted to be used as Road Transports viz Type 4 with driver or Type 5, driverless, the routes where these AVs can be permitted – long or short or in specific locations like airports, railway stations and ports.

5.5 Changes required to MV Act and Rules before permitting self-driving AVs to carry out testing or ply on Indian Roads

Analysis of the various legal provisions of the MV Act, Rules and Allied enactments helps to conclude that before permitting self-driving AVs to carry out testing or ply on Indian roads a minimum of following changes will be required to the MV Act (Act) and Central MV Rules (Rules):

(a) Section 2(9) of the Act defines "Driver" as a person who acts as a steersman of the vehicle[744]. This will have to be modified to include "Remote Drivers" in the case of self-driving AVs.

(b) Section 2(10) of the Act defines "driving license" as permission to use a motor vehicle of a certain class or type[745]. Self-Driving AVs are presently not defined in the MV Act, and they will have to be defined as a specific class under the MV Act.

(c) Section 2(28) of the Act defines "motor vehicle" or "vehicle". Broadly speaking this definition is adequate to cover an AV which runs on any fossil fuel but may not include self-driving AVs running on battery[746]. Hence it is recommended that this definition may be tweaked to include all categories of self-driving AVs.

(d) Section 7 of the Act deals with "Restrictions on the granting of learner's licences[747] for certain vehicles". As of date, there is no specific mention of self-driving AVs in this section. Once self-driving AVs are defined under Section 2(10), this section needs to be tweaked to include restrictions on granting learners' licences for self-driving AVs.

[744.] Id.
[745.] Id.
[746.] Id.
[747.] Id.

(e) Section 10 of the Act deals with "Form and contents of licences to drive". This section lists the type of licences[748] which can be endorsed as a learner license or driving license. Once Section 2(10) is modified to include self-driving AVs then that list must also be included in Section 10.

(f) Section 14 of the Act deals with the "Currency of licences to drive motor vehicles". This section prescribes the validity period[749] for learner's and driving licenses and the conditions for renewal at different ages. Should the government decide on a different renewal process for self-driving AV driver's licenses then this section will have to be amended.

(g) Section 41 of the Act deals with "Registration, how to be made". This section amongst many aspects on registration of vehicles also prescribes the validity of a vehicle registration, which as of now is fifteen years[750]. Should the government decide to reduce or increase the validity period of a self-driving AV, then the same should be accordingly annotated in this section through an amendment.

(h) Chapter XI of the Act focuses on Third Party Risk Insurance for Motor Vehicles. This chapter covers the insurance requirements[751] for any damage, loss and loss of life of third parties caused by the driven motor vehicle. In case of vehicles carrying dangerous or hazardous goods, the vehicle(s) must also have a policy of insurance as defined in the Public Liability Insurance Act 1991. As per the present norms, the vehicle owner must have this third-party insurance in place. However, looking at the way the law has developed in countries where AVs are plying or are being tested, where the law requires manufacturers also to have in place similar insurance during the testing and commercial sale phase of the vehicle as liability for AV accidents can also be attributed to a manufacturing defect. The manufacturer is then required to have back-to-back indemnity with their other vendors and software suppliers. Similar provisions and best practices must be adopted in the India MV Act as and when self-driving AVs are permitted to be tested and plied in India.

[748.] Id.
[749.] Id.
[750.] Id.
[751.] Id.

(i)   The First Schedule of the Act has mandatory signs[752] under MV Act. A sign needs to be added in this to indicate Entry Prohibition for Self-Driving AVs as there will be in future many crowded areas in India where possibly self-driving AVs may not be permitted. That board will come in handy at that time.

(j)   Rule 8A of the Rules prescribes Minimum training needed to operate an electric rickshaw or electric cart and Rule 9 of the Rules prescribes Educational Qualifications for drivers of cargo vehicles transporting hazardous or risky items[753]. In a similar fashion training and educational qualification requirements, as approved by the government, must be prescribed for drivers of self-driving AV cars whether they are seated in the car or controlling it remotely.

(k)   Rule 11 of the Rules pertains to Learner's license preliminary test and Rule 15 pertains to the driving test[754]. Both these rules have laid down the syllabus for testing concerning conventional motor vehicles. These rules will have to be tweaked for including a syllabus for self-driving AVs as and when they are introduced in India.

(l)   Rule 123 to Rule 127 pertains to Safety Devices for Drivers, Passengers and Road Users[755]. Apart from the many safety devices listed in these rules which would be needed for a self-driving AV, in line with recommendations and regulations in other countries, it would also be prudent to have a black box or recording facility in self-driving AVs. If this finds favour with the government then a new rule will have to be added between Rules 123 to 127.

5.6   Concept of Product Liability in India, its comparison with the rest of the world

The word "Product Liability" is heard of in most advanced countries. Product liability law gives customers legal redress for any harm they sustain as a result of utilising a defective product. Many people across the globe are directly or indirectly affected due to the usage of defective products. These finally end up

---

[752.] Id.

[753.] Id.

[754.] Id.

[755.] Id.

in suits, where manufacturers pay up huge sums of money as damages. Any product that a consumer uses is reasonably expected to cater to the needs of the consumer and there lies a great deal of responsibility with manufacturers and sellers towards the safety and quality of these products. But in practice, this does not always happen. During the 18th to 20th century the adage caveat emptor, which translates as let the buyer beware, did govern general consumer law. Consumers reached out directly to manufacturers for a defective product and no court or law was involved. Post 20th century beginning the production style changed, the supply chain evolved, international trade grew, and e-commerce started, and all these led to the development of product liability laws[756] to protect consumers from unethical and fraudulent practices. It is the manufacturer who now must be more aware of defective products and the reputational risks associated with their sale.

Product Liability for self-driving AVs is an evolving area as the product and technology is still under development and hence cannot be compared with other existing off-the-shelf products including conventional driver-driven automobiles. As and when self-driving AVs become more prominent across the world, other global concerns like terrorist attacks and cyber-attacks[757] will be a cause of concern and product liability can also hinge on algorithms or GPS data, apart from other elements that go into the self-driving AV. The landscape of the ever-evolving product liability law in India is discussed in the next few paras as that's an important aspect of law in any regime which permits AVs to ply.

The earlier Consumer Protection Act 1986 of India was recently replaced by the Consumer Protection Act 2019 (CPA 2019) which does have the concept of product liability in line with global product liability laws. Before CPA 2019 since there was no specific product liability clause in any Indian statute, the product liability was governed primarily by contracts, applicable clauses in the Consumer Protection Act of 1986, Sale of Goods Act of 1930, Indian Penal Code of 1860 (presently Bharatiya Nyaya Sanhita), and certain specific statutes relating to particular goods and standardization[758]. The CPA

[756.] Don Mayer et al., Basics of Product Liability, Sales and Contracts (2012).

[757.] Martin Kahl, Autonomous vehicles: Driving regulatory and liability challenges Automotive World (2020), https://www.automotiveworld.com/articles/autonomous-vehicles-driving-regulatory-and-liability-challenges/ (last visited September 1, 2021).

[758.] Anindya Ghosh, Product liability law in India: An evolution - consumer protection - india Welcome to Mondaq (2020), https://www.mondaq.com/india/dodd-frank-consumer-protection-act/974270/

2019 outlines precise obligations for a product seller, service provider, or manufacturer specifically for compensating consumers for harm caused by a defective product. The word "Consumer" is defined in section 2(7) which includes online consumers and "Product" has been defined in section 2(33) of CPA 2019[759].

As per CPA 2019 product is wide enough to not only cover goods, but can also include sources of supply or intermediary goods, but excludes human body parts. A person using a product for commerce is not a consumer[760]. Product liability under the CPA 2019 is essentially described as the seller's or manufacturer's obligation to make up for the harm caused to the customer by using a defective product, particularly regarding the product's quality and safety[761].

Chapter VI of the CPA 2019 covers Product Liability from Sections 82 to 87[762]. Section 82 makes Chapter VI applicable to every compensation claim for product liability on account of defective product manufactured, sold or serviced by the seller[763]. Section 83 is an enablement provision, which allows a complainant to bring about a product liability case for a defective product against a manufacturer, a supplier of goods or a retailer of goods[764]. Section 84 implies product liability on the manufacturer only if the product has a manufacturing/design defect or it deviates from specifications[765]. Being non-negligent in making an express warranty of the product is no defence for a manufacturer. Section 85 pertains to product service providers and makes them liable only if the service provided was faulty or imperfect or deficient or inadequate in quality[766]. If the service does not adhere to the express guarantee or the contract's terms and conditions, the product service provider is also accountable. If a non-manufacturer had a significant role in the designing, testing, manufacturing, packing, or labelling of a product that caused harm,

---

product-liability-law-in-india-an-evolution (last visited September 1, 2021).

[759]. Ibid

[760]. Rachit Garg - et al., Product liability and consumer protection iPleaders (2021), https://blog.ipleaders.in/product-liability-and-consumer-protection/ (last visited September 1, 2021).

[761]. Ibid

[762]. The Consumer Protection Act 2019 [online] available at https://egazette.nic.in/WriteReadData/2019/210422.pdf (last visited March 21, 2021)

[763]. Ibid

[764]. Rachit Garg - et al., Product Liability and Consumer Protection iPleaders, Supra Note 772, at 206.

[765]. Id.

[766]. Id.

Section 86 would extend the product liability obligation to them[767]. Also, a product seller whose alteration/modification of product or tall warranty claims caused harm can be made liable[768]. Section 87 lays down the exceptions for actions against the product seller and manufacturer for product liability. As per section 87, The product provider is not liable if the item was used improperly, changed, or altered when the damage occurred[769]. The producer of the product is not responsible for failing to inform or forewarn a customer about a risk that is evident or well-known[770].

5.7 Product Liability Case Laws

Winterbottom v. Wright (1842)[771]: Strict responsibility in tort law and negligence developed as an outcome of the well-known English case of Winterbottom *v. Wright (1842)*[772] which upheld the privity requirement. In this instance, a contract was made between a coach firm and the postmaster general to supply coaches for postal service and to assume maintenance duty for the aforementioned coaches. Due to inadequate maintenance, one of the coaches collapsed causing injuries to the plaintiff hired by the postmaster general, who then sued the coach company. The driver was not a party to the maintenance contract, so the court concluded that he could not file a lawsuit against the bus firm for damages.

MacPherson v. Buick Motor Co (1916)[773]: The second most important case is that of *MacPherson v. Buick Motor Co (1916)*. This case[774] abolished the necessity of contract privity for negligence. The background of this case is that the plaintiff, Donald C. MacPherson, got injured because of the collapse of a wooden wheel of his 1909 Buick Runabout. Buick Motor Company

---

[767]. Id.

[768]. Id.

[769]. Bindu Janardhanan Sidharth Sethi, PRODUCT LIABILITY UNDER THE CONSUMER PROTECTION ACT, 2019: LET THE MANUFACTURER/SELLER BEWARE! BAR AND BENCH - INDIAN LEGAL NEWS (2020), https://www.barandbench.com/columns/product-liability-under-the-consumer-protection-act-2019-let-the-manufacturer-seller-beware (last visited September 1, 2021).

[770]. Rachit Garg - et al., Product Liability and Consumer Protection iPleaders, Supra Note 772, at 206.

[771]. 10 M&W 109.

[772]. Winterbottom v Wright – 1842 [online] available at <https://www.lawteacher.net/cases/winterbottom-v-wright.php?vref=1> (last visited April 22, 2021

[773]. 217 N.Y. 382.

[774]. MacPherson vs Buick Motor Co [online] available at https://www.casebriefs.com/blog/law/torts/torts-keyed-to-prosser/duty-of-care/macpherson-v-buick-motor-co-2/ (last visited March 22, 2021)

was the defendant in this case as they were the manufacturer of the vehicle. However, the wheel was not manufactured by them and was made by a third party. Nevertheless, the defendant installed a wheel made by a third party, and the evidence in the case revealed that the fault in the wheel might have been found with reasonable inspection, which the defendant never performed. However, the defendant pleaded in court that they were not liable as the plaintiff had not purchased the vehicle from the defendant, but instead the purchase was made from a dealer. Judge Cardozo of the New York Court of Appeals in his ruling mentioned that since the company was negligent, they were liable and hence cannot take the defence of lack of privity of contract with the plaintiff. The idea of the privity of a contract had never before been rejected by a court until this one. As per legal luminaries, this was a victory of tort over a contract[775].

Escola v. Coca-Cola Bottling Co. (1944)[776]: In the olden days for proving that the defendant was negligent, a plaintiff had to prove that the defendant did not apply a required standard of care. This practice is very tough to prove and hence the 20th-century courts started moving in a different direction as they thought that it was unfair to force sufferers to support manufacturer liability claims for carelessness. While deciding the case of Escola v. Coca-Cola Bottling Co. (1944)[777], Justice Traynor expressed his views stating once a manufacturer has placed a product in the market without an inspection, knowing fully well that it is to be used and the product is proven to have a defect which causes human injury, then the manufacturer incurs an absolute liability.

Greenman v. Yuba Power Products, Inc[778] **and Henningsen v. Bloomfield Motors, Inc**[779]: Courts those days also were introspecting to see if there was a case for a manufacturer's express or implied warranty to the consumer. The res ipsa loquitur concept, which states that the item speaks for itself, was also enlarged by courts at that time to lessen the plaintiff's burden of proof. The two famous cases of those days were the case on strict liability imposed on manufacturers under the Greenman v. Yuba Power Products, Inc.,

---

[775]. Ibid.

[776]. 24 Cal.2d 453.

[777]. Escola vs Coca Cola Bottling Co [online] available at https://www.casebriefs.com/blog/law/torts/torts-keyed-to-epstein/products-liability/escola-v-coca-cola-bottling-co-of-fresno/ (last visited March 24, 2021)

[778]. (1963) 59 Cal.2d 57.

[779]. 32.N.J. 358.

and the issue of the implicit assurance of safety through the Henningsen v. Bloomfield Motors, Inc. case.

The *Greenman v. Yuba Power Products Inc* is a 1963 case[780] where Greenman, the complainant, bought the device known as shop smith. This device was a multifunctional instrument that could be used as a drill, saw, and wood lathe. The plaintiff purchased shop smith after reading a brochure on the product and seeing a demonstration by the retailer. He also procured requisite attachments to use shop smith as a lathe. The plaintiff had used the lathe successfully many times before it threw a wood piece and injured him seriously in his head. The plaintiff then sued the manufacturer and the retailer. The higher court affirmed the lower court's judgement and said that the plaintiff as a consumer had the right to sue the manufacturer on grounds of warranty breach, so long as the plaintiff was able to prove that he used the product in the right way and that the injury was due to manufacturing design defect, which the plaintiff was not aware and that this design defect had made the product unsafe.

The *Henningsen v. Bloomfield Motors, Inc* is a 1960 case[781] wherein ten days after the plaintiff brought the car from a dealership owned by the defendant, the car's steering malfunctioned resulting in an accident when the wife of the plaintiff was driving. As a result, the plaintiff filed a case against the manufacturer. The dealer defended his position by stating that the plaintiff's warranty agreement had exempted the defendant from any liability arising from personal injury. A careful reading of the warranty clause indicates that during the first 90 days or 4000 miles of use, the guarantee only covered the repair of defective parts. Nevertheless, the court awarded damages in this case to Henningsen. The court order mentioned that every sale carries with it an implied warranty for safety and this warranty is extended to anyone who uses the product irrespective of the fact that the person was the owner or not.

A.S. Mittal v. State of U.P. (1989)[782]**:** With the progressive thinking of the judiciary, product liability cases have been gradually holding manufacturers accountable for negligence, wherever the injury is caused due to

---

[780] The University of Chicago Law Review, 1971. Strict Products Liability to the Bystander: A Study in Common Law Determinism. 38(3), p625.

[781] Harvard Law Review, 1961. Sales. Implied Warranties. Implied Warranty of Merchantability Renders Manufacturer Liable to Buyer's Wife despite Disclaimer Clause and Absence of Privity of Contract. Henningsen v. Bloomfield Motors, Inc. (N. J. 1960). 74(3), p.630.

[782] 1989 AIR 1570

manufacturing defect, even if no contract existed between injured/consumer and manufacturer. In India, courts have largely dealt with product liability claims using the concepts of strict liability and negligence. Historically there are no provisions for manufacturer or seller responsibility in Indian statutes on account of defective/faulty products or services till the enactment of CPA 2019. The case of A.S. Mittal vs the State of U.P. was a case where mass eye surgery went wrong and caused irreparable damage to the eye of many who underwent the surgery. The doctors in their statements mentioned that they took all precautions and still the infection happened. The saline, which was used to irrigate the eyes, bought from a reliable manufacturer, was probably the cause or source of contamination. The court held that since was product was to be used in a mass surgery, there were two questions to be answered i.e., whether a safety and purity pre-test was required and if doing so would result in liability. The court held that answers to these two questions must be adduced through evidence and the decision of the court would be based on facts and evidence presented in the case[783].

Airbus Industrie v. Laura Howell Linton (1994)[784]: The is the case of an aircraft accident which occurred on 14 February 1990 wherein a scheduled commercial flight IC 605, with registration no VT-EPN, flying from Mumbai to Bangalore touched the ground 2300 feet short of scheduled touchdown point in the runway. 92 passengers and four crew members were killed, and the other 54 survivors suffered grave injuries[785]. On 12-2-1992, respondents 4 to 9 filed a lawsuit with the Mumbai High Court against Indian Airlines (IA) and Hindustan Aeronautics Ltd (HAL) requesting compensation in Pound Sterling equal to Rs. 5,95,24,989-56 ps. The lawsuit's plaintiffs were all British citizens with Indian ancestry. The case mentioned in the suit is that the accident was caused because of insufficient and improper training of the pilots and their wrong understanding of aircraft systems. The respondents also mentioned that they doubted if the aircraft was airworthy or not. HAL was alleged to have not provided immediate and timely firefighting and rescue services[786]. Since product liability law did not exist in India at that time, respondents simultaneously filed a compensation suit in Texas courts in the

---

[783.] 1989 AIR 1570.
[784.] ILR 1994 KAR 1370
[785.] Id.
[786.] Id.

USA. On 6th November 1992, in Mumbai City Civil Court, the appellant i.e., Airbus Industrie filed a lawsuit against respondents 4 through 9 as well as 10 and 11, seeking a declaration that the respondents cannot proceed in any court that is not an Indian court. The respondents answered back stating that given the absence of product liability rules in India, Texas courts were more appropriate[787]. The City Civil Judge took notice of the application for a temporary injunction and dismissed it. The Hon'ble High Court of Bombay's earlier judgement, which was upheld until 15 December 1992 to give time for an appeal, caused the learned City Civil Judge to postpone the aforementioned decision[788].

On November 21, 1992, Airbus filed lawsuit O.S. No. 7517 of 1992 before the City Civil Judge in Bangalore, asking for a decision that respondents 1 through 9 are not allowed to file a lawsuit against Airbus in any court other than Indian courts and that their case in Texas courts is not compliant with Indian law[789]. The trial judge issued an injunction that was in effect until April 1, 1993. The same was periodically extended, and by order dated 10-6-1993, I.A.No. 3 was dismissed after weighing the defence of respondents 1 to 9[790]. This order of dismissal was appealed before the Karnataka High Court and the same was finally set aside by the Hon'ble High Court. Although India does not have rigorous product liability laws, the Karnataka High Court rejected the respondents' argument and stated that it is inappropriate to say that parties will have no recourse in such situations[791]. The court referred to the case of Charan Lal Sahu v. Union of India (Bhopal Gas Disaster) wherein damages were awarded and further stated that outdated laws must be changed or new legislation must be passed to assist in such circumstances[792].

---

787. Id.
788. Id.
789. Id.
790. Id.
791. Id.
792. Id.

5.8 Pre-CPA 2019 Product Liability Case Laws in India with valid defence

EID (Parry) India Ltd. v. Baby Benjamin Tushara[793]**:** This was a case where a consumer brought a claim of a manufacturing defect in a toilet commode after thirty years of using that product. The court held that unless the consumer could prove some inherent defect in the product, normal wear and tear will not make the consumer entitled to compensation[794].

A. K. Roy and Another vs Voltas Limited[795]**:** This was a case where the consumer had purchased Voltas's air conditioner. After the customer complained to the firm about the loud noise the air conditioner was making, the company replaced it, and the new air conditioner was no longer making any noise. When the consumer claimed further damages, the court held that since the air conditioner was replaced with another one which was functioning as intended, the customer is not entitled to further remedies[796].

M/s. Solidaire India Ltd. v/s D.A. Mohan Kumar[797]: This was a case where the consumer Mr. Mohan Kumar filed a case for free replacement of a defective picture tube in his Solidaire television set. The television set had a warranty for one year and the picture tube failed after three years of usage. The District Consumer Forum's view was that even if a warranty of one year was given it does not mean the picture tube's quality must last for only a year. The District Consumer Court had decided that the picture tube had a manufacturing defect without having it technically examined and this was overturned in appeal. Nevertheless, a small relief in terms of a discounted price for the replacement of the picture tube was given to the consumer by the Kerala State Consumer Disputes Redressal Commission Thiruvananthapuram[798].

Atul Anant Rane and Ors. Vs. Lourdes Gas Service and Ors[799]**:** This was a case filed with Maharashtra State Consumer Disputes Redressal Commission, Mumbai, by the children of Anant Subhash Rane and Sunita

---

[793] 1992 (23) DRJ 394

[794] Id.

[795] 1973 AIR 225

[796] Id.

[797] Appeal No. 18 of 1990, Decided On, June 29, 1991 at, Kerala State Consumer Disputes Redressal Commission Thiruvananthapuram - Unreported.

[798] M/s. Solidaire India Ltd. v D.A. Mohan Kumar on June 29, 1991, (1999), https://www.lawyerservices.in/Ms-Solidaire-India-Ltd-Versus-DA-Mohan-Kumar-1991-06-29 (last visited September 1, 2021).

[799] Complaint Nos. 34 & 288 of 1993, Decided on November 4, 1997 at, Maharashtra State Consumer Disputes Redressal Commission Mumbai

Anant Rane for their death due to gas leakage on 21 December 1991. The consumer forum held that the death was caused due to wrong handling of the gas knob and hence no relief was admissible[800].

5.9 Gaps to be plugged under Product Liability Laws in India to permit Autonomous Vehicles to be tested/plied on public roads in India

While product liability laws are existent in many countries of the world where self-driving cars are plying/being tested, these laws are not causing any concern to the owners/ drivers of these AVs. The lure of reduced crashes and connected lower insurance costs are luring drivers and owners to adopt AV technologies. Propounders of AV technology for self-driving cars feel that later the adoption of AV technology, the liability of manufacturers would increase manifold. For any AV crash, one can have two kinds of litigation, a no-fault insurance claim or a product liability claim. The reassignment of liability from the driver to the manufacturer in the case of AVs has made no-fault insurance more attractive than a product liability suit which is more expensive and time-consuming. No-fault liability traditionally provides quick resolution and relief in case of accidents as they are not dependent upon identifying who is at fault[801]. Globally product liability law is generally a hybrid of contract and tort law making the manufacturer liable for their products. The manufacturers are expected to have back-to-back liability clauses with all their vendors and in the case of AVs also same will be applicable. When it comes to product liability the courts will try to look for a single neck to choke viz the manufacturer, unless the AV software and other elements are sold separately by non-vehicle manufacturers, in which case they would be liable for their products. Wherever we talk of "Strict Product Liability", it would mean that there is no need to prove negligence[802]. Product

---

[800.] 1998(4) ALL MR (Journal) 20, Atul Anant Rane & Ors. vs. Lourdes Gas Service & Ors., NEARLAW. COM (1997), https://nearlaw.com/PDF/MumbaiHC/1998/1998(4)-ALL-MR-(JOURNAL)-20.html (last visited September 1, 2021).

[801.] James M Anderson & Nidhi Kalra, Liability Implications of Autonomous Vehicle Technology JSTOR (2014), https://www.jstor.org/stable/10.7249/j.ctt5hhwgz.14?seq=1#metadata_info_tab_contents (last visited September 1, 2021).

[802.] John Villasenor, Products liability and driverless cars: Issues and guiding principles for legislation Brookings (2018), https://www.brookings.edu/research/products-liability-and-driverless-cars-issues-and-guiding-principles-for-legislation/ (last visited October 1, 2021).

liability case rules and doctrine from the past did not differentiate between different types of faults. However, throughout time, three different types of flaws namely manufacturing flaws, design flaws, and warning flaws (failure to warn) have evolved[803]. All three are self-explanatory words and carry the same meaning that they imply.

In the Indian context CPA, 2019 would govern the aspect of Product Liability. As discussed earlier criminal liability and damages associated with criminal liability and product liability are two different areas. Product liability would be something which the seller/manufacturer is liable to pay up and in case they have multiple vendors to manufacture a product then they must have back-to-back liability contracts with these vendors to take care of any eventuality wherein they have to pay up damages on account of product liability. In the case of AVs, the manufacturer needs to have such back-to-back arrangements with all vendors including but not limited to vendors for various vehicle parts, sensor vendors and software vendors. Criminal liability would be governed by individual culpability and therein a case could be filed separately against the manufacturer, sensor manufacturer, radar manufacturer and software vendors in the case of an AV.

As of date product liability laws are only in one category and there are no distinct subcategories to take care of self-driving AVs. Under the current legal provisions claims concerning production, flaws would require that the proof of failure or manufacturing defect be proved, and a correlation established between the defect and the cause of the accident of the self-driving system. When dealing with product liability issues of self-driving AVs one must fully understand various components like radar sensors, laser rangefinders, video cameras, global positioning, or digital mapping, in an AV and how they interface with each other[804]. If it finally comes out that the error is only a software error then technically under the present interpretation, the software may not come under a manufactured product in most product liability laws and hence product liability cannot be claimed in most countries.

[803.] JAMES M. ANDERSON, Autonomous Vehicle Technology: A guide for policymakers (2016).
[804.] Cohen IADC product liability May 2015 - pbnlaw.com, (2015), https://pbnlaw.com/media/540169/Cohen-IADC-Product-Liability-May-2015.pdf (last visited October 1, 2021).

Because of the foregoing, CPA 2019 is as on date adequate when it comes to handling product liability cases that might arise due to the usage of self-driving AVs. To handle criminal liability for AV accidents, there are tweaks required to be made to the MV Act as discussed earlier.

5.10 Gaps in the Indian Evidence Act and Information Technology Act which can hinder operations of self-driving AVs in India

The more we have autonomous cars, the more the danger of these cars being constantly under cyber threats like hacking. If we look at the Indian Information Technology Act 2000 (IT Act 2000), as per Section 66(1) hacking with Computer Systems is defined as:

> "Whoever with the intent of causing or knowing that is likely to cause wrongful loss or damage to the public or any person destroys or deletes or alters any information residing in a computer resource or diminishes its value or utility or affects it injuriously by any means, commits hacking"[805] (Section 66 (1) IT Act 2000).

However, it can be successfully argued that the hacking of AVs will not count as hacking under the IT Act 2000. This is because the definition of computer resource in the IT Act 2000, does not include AVs. This is a clear gap and will have to be plugged as and when self-driving AVs are introduced in India for testing and plying. Apart from correcting the IT Act 2000, it is also required that the scope of hacking under IT Act 200 be widened, and strict provisions are incorporated in the said act to ensure that car manufacturers incorporate good anti-hacking mechanisms and safety provisions in AVs[806]. The IT Act 2000 must also be amended to include the issue of liability and insurance on account of accidents caused due to hacking. The Indian Evidence Act which now is revised as Bharatiya Sakshya Adhiniyam, as of date appears to be comprehensive enough to handle any case connected with AVs.

---

[805] The Information Technology Act 2000, Sec 66(1).

[806] The RMLNLU Law Review Blog, MOVING TOWARDS A DRIVERLESS ERA THE RMLNLU LAW REVIEW BLOG (2021), https://rmlnlulawreview.com/2017/10/25/moving-towards-a-driverless-era/ (last visited October 1, 2021).

5.11   ISO 21434:2021

It has been assessed that an ordinary adult in the US may produce user data worth about \$35 per month, or \$420 annually[807]. Since connected cars or AVs would have abilities to set cookies on your infotainment device and track your surfing habits, such as a smartphone, there is a fair possibility that the data produced by an AV would be of similar value[808]. ISO has recently published ISO 21434:2021 standards on cybersecurity engineering on-road vehicles. The design, creation, production, use, maintenance, and decommissioning of electrical and electronic (E / E) systems in motor vehicles, as well as those systems' components and interconnections, are all specified in ISO 21434 as standards for cybersecurity risk management. ISO 21434 also brings about a common language in cybersecurity for E/E systems throughout the supply chain. ISO 21434 in the long run will help manufacturers and their supply chains to be ahead of the curve in protecting their components and road users from cyber-attacks[809].

5.12   Conclusion

India has a huge demand for automobiles and the automobile industry is one of the largest in the world. However, the global automobile players do not see India as a potential hub for AVs soon due to various factors mentioned in the next few lines. As mentioned earlier, one of the major issues that the Indian Government sees with the introduction of AVs is potential job loss. Another big problem in India is the lack of proper infrastructure due to potholes and untarred roads. This makes navigation difficult for AVs. Lack of road discipline is another issue in India which is a potential damper to introducing AVs. Indian drivers are used to driving on the wrong side, people walk all across the roads, poor lane discipline, speeding and jumping signals are a norm of life in India, which has to change if we have to introduce AVs. Apart from this,

---

[807.] How much is user data worth? [online] available at https://pawtocol.medium.com/how-much-is-user-data-worth-f2b1b0432136 (last visited October 21, 2022)

[808.] Abhilasha Singh, Internet-connected cars: What data is collected, who has the access and more The Financial Express (2021), https://www.financialexpress.com/auto/car-news/internet-connected-cars-what-data-is-collected-access-hyundai-venue-kia-sonet-seltos-mg-hector/2305516/ (last visited October 1, 2021).

[809.] Rajeev Panicker, ISO 21434:2021 - an overview of newly published standard on Cybersecurity Engineering for Road Vehicles LinkedIn (2021), https://www.linkedin.com/pulse/iso-214342021-overview-newly-published-standard-road-panicker/ (last visited October 1, 2021).

the free movement of cattle and animal-pulled carts on roads, especially on highways is also something that has to be sorted out before introducing AVs in India.

Coming to the regulatory landscape in India, the MV Act and its amendments do not permit plying of AVs including carrying out road testing. Around four years back an amendment was mooted to the MV Act to permit AV testing and this proposed amendment has not gained any traction. While the ideal situation would be to have separate laws for AVs in India, that's not a practical solution. Hence a suitable amendment to existing laws concerning AVs, like MV Act, CPA 2019 and allied acts connected with motor vehicles should be good enough. The biggest question on AV accidents globally is insurance and who is liable i.e., owner or driver or manufacturer? MV Act/CPA 2019 have to be tweaked to cater for all these situations based on global precedence, including cases related to contributory negligence. Other important areas concerning AVs are data protection and privacy. Although there is no specific data protection law in India., the existing provisions of Sections 43, 65 and 66 of the Information Technology Act 2000 and the rules under Information Technology (Reasonable Security Practices and Procedures and Sensitive Personal Data or Information) Rules, 2011 would be applicable for all cases involving data breach/data protection concerning AVs.

It is quite clear from the analysis in this chapter and the preceding chapters that for introducing AVs in India, amendments will have to be made to MV Act and allied enactments as proposed earlier in this chapter. In addition to the above, we need to have a clear-cut law which captures the roles, responsibilities and liabilities of various actors involved in an AV like the driver, manufacturer, owner, software developer and third party (e.g., hacker). We also need strict laws to implement traffic discipline on our roads and manage cattle plying on roads. Strict and deterrent punishments for AV hackers are a necessity. One can safely conclude that self-driving AVs will be a part and parcel of human life going forward. The usage of these vehicles is going to be more in the commercial and public transport space for sure. Also, the use of AVs on EV platforms is likely to happen in foreseeable future. Product liability and regulations for self-driving AVs is likely to evolve gradually over a period. To have self-driving AVs ply in India, there are a lot of legal and infrastructural hurdles that will have to be navigated[810]. In the Indian context, while self-driving AVs can be a dystopian nightmare, they will help us experience a better quality of driving, ensure safe driving, and reduce traffic congestion and driving stress. However, the fear of sensors failing especially when driving in congested areas in India can be stressful. Machine Learning can help AVs learn to negate errors. Loss of jobs due to the introduction of AVs can be countered by the additional jobs which will be created in the IT / ITES space due to the introduction of AVs.

Product Liability which has been introduced in CPA 2019 is expected to mature in the years to come and one will see a lot of judgements in that area which will act as precedent going forward. It will also be interesting to see how the industry in India deals with and reacts to these changes to product liability laws and their transformation towards becoming more ethical and customer-friendly companies. Adequate data protection and privacy safeguards must also be provided to all users of AVs, to cover gaps in areas not covered under general data protection laws[811]. The introduction of AVs will help reduce road accidents and hence if public use of AVs is an issue or there is likely to be a delay, the ideal course would be to facilitate AV use in public transportation like shuttle services etc and then slowly move it to private space.

✳ ✳ ✳

---

[810] Autonomous Driving Technology: Can self-driving cars run on Indian Roads, Moneycontrol (2020), https://www.moneycontrol.com/news/driverless-cars/ (last visited October 1, 2021).

[811] Diganth Raj Sehgal - et al., Legal issues related to Autonomous Vehicles iPleaders (2020), https://blog.ipleaders.in/legal-issues-related-autonomous-vehicles/ (last visited October 17, 2021).

CHAPTER 6

CONCLUSION, FINDINGS AND SUGGESTIONS

6.1 Preliminary

A self-driving AV is a vehicle which can move on its own with the help of its various sensors. These sensors allow the AV to sense the surrounding environment of the vehicle and move without causing an accident or damage. A fully self-driving AV needs no physical driver inside the vehicle, nor does it require a physically present human passenger to take control. These AVs can go anywhere a normal car can go and do everything that an experienced and capable driver can do. Six degrees of automated driving have been established by the SAE, ranging from Level Zero, which is manual, to Level 5, which is fully autonomous. The SAE levels are a standard that has been adopted by USDOT and other countries across the world are following similar guidelines.

The words "autonomous", "automated" and "self-driving" all look similar but have subtle differences. The dictionary definition of "autonomous" is the capacity for self-administration. "Automated" refers to being run primarily by automatic machinery. and "self-driving" means a machine manoeuvred by a computer and operating without human guidance. Autonomous cars work on various kinds of sensors, and complex computer programs and use powerful processors for executing the software[812]. The sensors in these vehicles along

---

[812] What is an autonomous car? – how self-driving cars work, SYNOPSYS, https://www.synopsys.com/automotive/what-is-autonomous-car.html (last visited December 12, 2021).

with the software programs help these vehicles to create a virtual map of their surroundings and thereafter intelligently steer the vehicle to move smoothly and avoid any collision. Some of the hardware elements used by these AVs are radar, lidar, video cameras and different types of sensors. The video cameras in these vehicles help to detect traffic signals, decipher road signs, and track pedestrians and other vehicles. Lidar stands for light detection and ranging. The lidar sensors bounce light pulses from the AV to objects in the surroundings for measuring distances. Lidar also helps detect the edges of the road and identify the lane markings. The AV wheels also have ultrasonic sensors, and these sensors help to detect curbs and other obstructions including vehicles when parking. The software in the AVs helps process the sensory input by plotting a path to move and thereafter controls the car's speed, brakes, and steering wheel. The software helps the AV to avoid obstacles and follow traffic rules.

6.2   Advantages of Autonomous Vehicles

As per the National Safety Council of the USA, approximately 40,000 persons die due to road accidents and the number is increasing year after year. Apart from these deaths, millions have also been injured through road accidents and their medical treatment has been costing millions of dollars to the US exchequer. As per the NHTSA study since human error is the main cause of road accidents, having self-driving AVs ply on the road is the best way to reduce accidents, deaths, and medical treatment costs. Insurance companies in the USA are also supporting this, and they have started offering discounts to drivers of vehicles with safety devices. Some of the popular benefits[813] of self-driving AVs can be summarised as a reduction in accidents by 90%, a drop in harmful emissions by 60%, elimination of waves of traffic which are created through stop and move human behaviour, improvement of fuel economy by 10%, reduction in travel time by 40% and savings by consumers to the tune of 5 billion pounds.

[813.] 7 benefits of Autonomous Cars, THALES GROUP (2017), https://www.thalesgroup.com/en/markets/digital-identity-and-security/iot/magazine/7-benefits-autonomous-cars (last visited December 12, 2021).

6.3 Legal regime for AVs Globally

The fundamental premise of any AV Policy is that automated vehicles can help reduce accidents through better safety, sustainability, and mobility. US DOT has specific guidelines[814] for manufacturers involved in the development and testing of vehicles. The NHTSA published its first version of the Federal Automated Vehicles Policy or FAVP in September 2016. Section 2 of the Model State Policy or MSP clarifies the division of federal vs state responsibilities[815]. As of 30 April 2022[816], thirty-eight states along with the District of Columbia (DC) have passed laws/executive orders for testing and plying of Autonomous Vehicles (AVs)[817]. Five states have only authorized studies to define the framework for AVs and to look into various funding aspects for the research, development and infrastructure related to plying of AVs. Twelve states have permitted testing. Sixteen states and DC have permitted the full deployment of AVs. Amongst the above, testing and deployment without a safety driver are permitted in 18 states and truck platooning is regulated in four states. Other states either have them in the pipeline or are yet to decide on them. AVs in the USA are rated according to their SAE automation levels.

The "Safely Ensuring Lives Future Deployment and Research in Vehicle Evolution Act" (SELF DRIVE Act), also known as H.R. 3388, was enacted by the US House of Representatives in 2017. This bill aimed to help adopt federal safety standards in the arena of AVs. The bill attempted to clarify the roles of federal and state governments regarding this emerging technology. Post passing of this Act, a similar bill called "American Vision for Safer Transportation through Advancement of Revolutionary Technologies Act" (AV Start Act) under reference S. 1885 was also presented but never passed. Both the AV Start Act and the SELF-DRIVE Act's draft language call for the creation of a "Highly Automated Vehicle Advisory Council" within the Department of Transportation to assess self-driving car-related issues and resolve federal regulation surrounding AV testing. Under the AV Start Act, the onus was vested in each AV manufacturer to provide a defined strategy for assessing

---

[814.] Fact sheet: Federal automated vehicles policy overview, https://www.transportation.gov/sites/dot.gov/files/docs/DOT_AV_Policy.pdf (last visited January 10, 2021).

[815.] Anne Teigen Ben Husch, Regulating Autonomous Vehicles, https://www.ncsl.org/research/transportation/regulating-autonomous-vehicles.aspx (last visited January 10, 2021).

[816.] Who is responsible in case of accidents involving a driverless car? [online] available at https://www.americanbazaaronline.com/2022/05/06/who-is-responsible-in-case-of-accidents-involving-a-driverless-car-449512/ (last visited July 10, 2022).

[817.] Autonomous Vehicles, GHSA, https://www.ghsa.org/taxonomy/term/521 (last visited January 10, 2021).

and minimising cybersecurity risks. The Department of Commerce was then responsible to set up a committee for data access which would make appropriate policy recommendations concerning data production and sharing. The Self-Drive Act clarified the role of NHTSA in regulating the safety, design and construction of AVs. It improved the access of NHTSA to safety data to help NHTSA plan any required upgrades and research and development (R&D). In addition, the Self-Drive Act also created manufacturer certifications for NHTSA safety assessments and catered for enhanced protection for consumer education, privacy, and cybersecurity.

To hasten the market introduction of these vehicles, the Self-drive Act also proposed to exclude the first 100,000 vehicles produced by a manufacturer from already-existing safety regulations[818]. The main objection to the Self-Drive Act was the absence of regulatory or testing requirements. Experts felt that the Self-Drive Act avoids holding technology and automakers accountable when their unregulated goods put customers at peril. Finally, both bills came to an end when the new Congress took office in 2018. Two Republican Senators have reintroduced[819] the Self-Drive Act again on September 23, 2020. The goal, according to these senators, is to establish a national framework for the research and testing of AVs[820].

Manufacturers and AI technology innovators in Europe are working rapidly on self-driving AVs. Many countries in Europe with the support of their government are working hard to become an AV leader. A lot of research and trials on AVs are being undertaken across many countries in Europe. These countries are also trying to align their AV programs with other countries for uniformity and traction. Germany which has many world-famous automobile companies like Volkswagen, Mercedes and BMW has also been the pioneer nation in Europe to come up with an Autonomous Vehicle or AV policy. In the year 2017 Germany enacted a law which permitted for testing and use of AVs on public roads. One of the mandatory requirements in German law was to have a safety driver in the car who could control the car and the steering

---

[818.] H.R. 3388 (115th): Self Drive Act, GovTrack.us, https://www.govtrack.us/congress/bills/115/hr3388/summary (last visited December 17, 2020).

[819.] Text - H.R.8350 - 116th congress (2019-2020): Self Drive, https://www.congress.gov/bill/116th-congress/house-bill/8350/text (last visited December 22, 2021).

[820.] Republican House Members Reintroduce Self Drive Act, AASHTO Journal (2020), https://aashtojournal.org/2020/09/25/republican-house-members-reintroduce-self-drive-act/ (last visited December 17, 2020).

wheel on a required basis. The present German legislation does not legally recognize level five or full AVs.

During the year 2017, an Ethics Committee was formulated in Germany to study and make recommendations on ethical programming in AVs. This committee came up with a report containing 20 different principles. Some of these principles were:

(a) Using networked AVs is ethically necessary only if they cause fewer accidents than human drivers.
(b) If there is an imminent danger, then the protection of human life must take priority.
(c) Qualification of people as per their characteristics like sex and age should not be done in cases of unavoidable accidents.
(d) When it comes to AVs, basis the driving conditions and situations the law must clearly define the parameters to identify the person/machine responsible for the driving task. This will kill any kind of ambiguity.
(e) Any personal information collected from the AVs must be done with the consent of the safety driver and they must have a say in what data can be collected.

The Scottish Law Commission and the Law Commission of England have been given mandates by the UK government in 2018 to review for three years, all laws related to driving, for identifying obstacles which could impede the introduction of AVs and possibly need legal reforms. The government mandate to both the Law Commissions revolves around the following:

(a) Can the driver be earmarked as a responsible person in an AV?
(b) If the AV is jointly being controlled by humans and machines, what should be the process to decide on civil and criminal liability/responsibility?
(c) How effectively can AVs be used in public transport areas?
(d) What is the forecasted demand or likely demand for driverless passenger transport?
(e) Can AVs be car shared through MaaS (Mobility as a Service) which is one of the emerging business models?
(f) Is there a need to come up with new kinds of criminal offences to deal with different types of conduct that may come up with the use of AVs?
(g) Impact of using AVs on road users and ways and means to protect road users from associated risk.

The European Commission released a document around May 2018 which had the heading "On the road to automated mobility: An EU strategy for future mobility"[821]. This document listed out a roadmap to outline the ambition of the EU towards becoming a world leader in deploying AVs. The Commission was of the firm view that the existing EU legislation was broadly conducive to permitting AVs to be put on the road and that only some changes were required in the law to synchronize the plying of AVs and non-AVs as a process towards making the law future-proof. The communication released by the EU Commission also talked about the following focus areas:

(a) Allocation of budget and investments towards the development of technology and requisite infrastructure for the operation of AVs.

(b) Development of demand in local markets to accept the usage of AVs.

(c) Issuance of detailed guidelines for plying of AVs including various road-related safety aspects.

(d) Issuance of detailed guidelines containing detailed requisite safety features for AVs.

(e) Liability aspects are connected with accidents and cover cybersecurity, data protection and access.

(f) Study on impact and implications of AVs on the society and economy as a whole and analyse the negative impacts if any needing amendments to existing laws or the introduction of new laws.

The Commission has also requested member states to finalise and prioritize their list of use cases for large-scale testing of AVs. These countries have also been advised to be mindful of likely synergies between connectivity and automation use cases. The Commission has also promised to ensure the availability of a common platform in the EU where all relevant public and private stakeholders will be mapped to facilitate each one of them to coordinate road testing of AVs and discuss and work towards solving pre-deployment issues.

Autonomous vehicles or AVs are rapidly growing in the automobile sector arena in Asia Pacific Japan China (APJC) region due to the great amount of time and investment being spent by most nations towards Research and

---

[821.] Ec.europa.eu. 2021. [online], https://ec.europa.eu/transport/sites/transport/files/3rd-mobility-pack/com20180283_en.pdf (last visited March 02, 2021).

Development (R&D) in this area. It is estimated that by 2024, APJC will probably have 24 million AVs. Apart from AVs another big thing on which nations are focusing is climate change and towards this Electric Vehicles or EVs are also an important aspect of R&D[822]. Many nations across the world are working towards integrating AVs and EVs. In the APJC region, many countries are focusing towards building appropriate infrastructure for EVs, which could propel the growth of AVs too. Singapore and Malaysia are pioneer nations in this regard.

Many automobile manufacturers in APJC and ANZ are mutually collaborating amongst themselves to control costs and bring about better efficiency. Some of the major players are Volvo, Ford, Honda, General Motors, Volkswagen, and Toyota. Apart from these companies, Uber is also involved as a shared mobility solution provider. To create new AVs, Honda has a deal with General Motors and its subsidiary Cruise. The planned investment is $2 billion in 12 years. As a beginning, the JV has made an immediate investment of $750 million. Shanghai has created a 100 square kilometres pilot zone with a lot of sensors and communication modules to facilitate the testing of AVs. Some of the brands which participated in the first round of pilot testing are Ford, Volvo, Cadillac and Volkswagen. China's 2025 vision includes Intelligent Connected Cars as a key strategic priority. As per Chinese experts, AV driving in China is more complicated than in the USA and hence if AVs achieve success in China, they can easily ply in the USA.

A two-and-a-half-square-mile self-driving taxi service in Singapore has been introduced by the MIT-based business NuTonomy. Toyota from Japan has also been carrying out AV trials on the Shuto Expressway in Tokyo. Toyota e-Palette was used to ferry athletes at the 2021 Japan Olympics and the Tokyo Paralympic games[823]. During the Tokyo Paralympic games after one such AV collided with a visually impaired athlete, the running of self-driving AVs was stopped[824]. Bosch in Australia, in 2016 did Joint Venture (JV) with the Government of Victoria to showcase an AV built using components from

---

[822] Autonomous cars to grow to 24 million units in Asia-Pacific by 2024, Supra Note 340, at 92.

[823] HT Auto Desk, Indian contingent at Tokyo Olympics to ride in these electric driverless cars Hindustan Times Auto News (2021), https://auto.hindustantimes.com/auto/news/indian-contingent-at-tokyo-olympics-to-ride-in-these-electric-driverless-cars-41626768980716.html (last visited December 22, 2021).

[824] Scooter Doll & Scooter Doll Scooter Doll is a writer, Toyota suspends all self-driving vehicles at Paralympic Games after collision with athlete Electrek (2021), https://electrek.co/2021/08/27/

Bosch and a Tesla body. Australian roads are famous for Kangaroo accidents. Every year more than twenty thousand Kangaroos are killed in road accidents in Australia and the insurance claims have always been beyond $Seventy-five million. Volvo has been studying this aspect specifically to help them build the right kind of software for AVs, to help detect kangaroos on the road and avoid colliding with them.

6.4   Analysis of AV laws in various Jurisdictions

The USA has been a pioneer when it comes to AVs in terms of infrastructure and AV laws. Also, many countries which are adopting AVs are following the USA AV testing policy including insurance for AV testing. A deep dive analysis of AV laws in various states of the USA has revealed the following:

(a) ADS must operate the AV safely even if there is a partial systems failure

(b) All vehicles must mandatorily have recording devices to cater for legal evidence

(c) All vehicles must be mandatorily covered by insurance to cater for liability claims. When responding to damage claims, evidence of financial responsibility must be shown in the absence of insurance coverage.

(d) Infrastructure like satellites, road striping, interactive signs and traffic signals are mandatory to help vehicles figure out when and where to adjust their vehicle speed and how to change lanes or allow other vehicles to change lanes

(e) Plying of truck platooning systems must be done only after permission from appropriate authorities

(f) Speeds for testing of AVs are specified in some states and there is also an ask for huge surety bonds to be submitted before testing e.g the amount is $5 million in California

(g) Maintenance of safe distance by testing vehicles and by convoys.

(h) Maintenance of federal safety standards by AVs

(i) Submission of yearly test reports by DOT to the designated committee in the state government

(j) Compliance with NHTSA standards for AVs

---

toyota-suspends-all-self-driving-vehicles-at-paralympic-games-after-collision-with-athlete/ (last visited December 22, 2021).

(k) All AV drivers whether in a vehicle or remote, to have a valid driving license. In the case of the remote driver, physical presence is to be in the USA.

(l) All AVs which are being tested must have permits for testing.

(m) An ADS permitted to operate without a driver, shall when engaged be considered a driver to adhere to the relevant motor vehicle laws.

(n) When automated vehicle repairs are carried out following manufacturer requirements, auto technicians and repair shops are excluded from product liability.

(o) Immunity to the manufacturer of an AV whose vehicle was converted as AV by a third party

(p) Autonomous vehicles before being permitted to be tested must have

   i.   a way to activate and deactivate autonomous technologies,
   ii. a visible sign outside the car indicating the use of the AV technology,
   iii. a device that can warn the safety driver to take over physical control of the autonomous vehicle if a failure occurs or is predicted to occur.

(q) When a youngster under the age of twelve is travelling in an autonomous vehicle, an adult must be present.

(r) In the case of a critical equipment failure, another emergency, or other circumstances necessitating a minimal risk condition, an autonomous vehicle must be able to do so automatically.

(s) Data collected by AVs may be shared for the goals of enhancing traffic control, security, or vehicle safety, if the owner, occupants, or human driver's identity is kept a secret.

(t) While the AV is moving, the safety driver can watch a video display that was placed at the factory from the front seat.

Europe has been closely following the USA when it comes to AVs. Analysis of the salient features/clauses in the AV laws/bills of various countries in Europe brings out the following common provisions in most AVs laws across Europe:

(a) Speed restrictions while testing AVs

(b) Autopilot with an autonomous change of lanes to be engaged only on highways/freeways.

(c) Fully automated military vehicles which have run a minimum of 300 kilometres can be put to the test on open roads with a backup driver ready to take over.

(d) All AV testing must have adequate Insurance coverage.

(e) Approval for testing from the appropriate Ministry in the government. Application for permission must specify the level of automation.

(f) Vehicle operators in the street where testing is to be carried out must be forewarned and adequately informed about the test.

(g) A safety driver who can take over control of the vehicle when necessary must be housed inside AVs.

(h) Since AV can detect white lines better than yellow lines, approval for painting yellow lines as white.

(i) The automated system must ensure the availability of adequate reaction time for the safety driver to take control of the AV.

(j) Rules outlining the safety driver's obligations and rights when using an AV in fully autonomous mode.

(k) Provisions regarding data and data handling like mandatory storage period, handing of data to police/enforcement authorities and data retention period. The video retention period in Russia is ten years.

(l) Whenever AVs are tested on public roads, they must have in place appropriate insurance.

(m) AV manufacturers and research centres must certify that the AV is ready in all aspects for road testing in the place where it is to be tested after considering all infrastructure in place.

(n) AVs to mandatorily have a data recording system (Russia) in Russia, to identify a vehicle as an AV, it must have a large black letter A on it.

Many countries in APJC and ANZ do not have AV-specific policies/laws in place. However, analysis of existing AV laws in countries where it exists and the scope of the existing normal motor vehicle laws for AVs, if made applicable brings out the following:

(a) Proposed PRC National Standards for AVs classify AVs on similar lines like SAE levels

(b) The person who applies for the testing AVs in PRC must have a registered legal status in PRC

(c) Qualifications prescribed for AV drivers

(d) AV is being tested to have prescribed insurance. In PRC it is RMB 5 million per tested vehicle and S $ 1.5 million in Singapore (PRC, Singapore)

(e) Data is to be stored in-country unless permission and consent are taken to store it out of the country.

(f) Revised safety standards laid down for AVs (Japan).

(g) Specific approval is required for the installation of cameras and sensors as per the revised motor vehicle act in Japan.

(h) AV manufacturers are required to provide all technical information about their vehicles to technicians who are repairing or servicing their vehicles (Japan)

(i) AV trials are to be carried out only after permission from appropriate authorities (Singapore)

(j) Level 3 AVs are only permitted for driving in Korea.

Analysis of liability-specific clauses for AVs in various European countries reveals the following:

(a) Applicability of general/tortious liability rules for any damage caused by AV or semi-AV (Austria, Italy. Monaco, Portuguese).

(b) The car operator/Owner would be responsible for any harm the vehicle caused unless the car was used without the operator's consent (Austria, Italy, Portugal, Romania). Exemption from liability during force majeure (Poland, Romania). The owner of a motor vehicle is liable for any damage brought on by the operation of that vehicle, whether it is operated by him or someone else. This is basically "No Fault Liability in Swiss parlance. Vehicle operators are liable under a fault-based regime in Switzerland.

(c) Product liability also applies to software that's part of a physical product like an AV (Austria).

(d) Warranty claims can only be made against the seller and not against the manufacturer of the AV (Austria).

(e) In the event of damages caused by automobile accidents, a strict liability rule may be imposed (In most countries in Europe including Germany).

(f) An AV's owner is responsible for any damages caused by his vehicle as per existing provisions of the StVG (Germany).

(g) The manufacturer and/or the AV software developer would be covered by current general strict liability laws for harms brought on by the AV in "autonomous mode". The only exception would be if it is proved that the damage was unavoidable (Hungary, Italy).

(h) The manufacturer can be held liable for product liability under the general liability laws and could be made liable under liability damages for hazardous products, which are moveable goods and carry strict liability regardless of fault (Poland, Ukraine).

The most prominent difference between the US regulations and that of Germany is in the definition of AVs. In the USA Level 5 AVs can be driverless. But as per StVG in Germany, there is a need to have a safety driver in all AVs and they must have the capability to take over control on a required basis.

Analysis of liability-specific clauses for AVs in APJC and ANZ countries reveals the following:

(a) When there are two motor vehicles involved in an accident, the party at fault is responsible and where the fault is on both sides the calculation of liability shall be on a proportionate basis of the fault (PRC)

(b) For accidents between an automobile and a pedestrian, the liability is always on the driver, unless the driver can prove that he/she was not at fault. If there is evidence that the pedestrian/non-vehicle was also at fault, then the liability is to be proportionately allocated based on the degree of fault. If it is proved that the driver was not at fault, then his compensation liability shall not exceed 10%. If it is proved that the accident occurred purely due to the fault of the non-vehicle/pedestrian then the driver shall not be liable for any compensation (PRC)

(c) Mandatory Third-Party insurance required in PRC

(d) Product liability as per terms of sales contract (PRC)

(e) Driver and Manufacturer's liability as per existing laws (Singapore)

(f) The law on the Protection of Consumers' Rights prescribes the right guidance (product liability) to calculate the liability for damages (Vietnam).

Legal Liability is another major issue to be addressed in APJC and ANZ for the smooth use of AVs. For the present manually driven vehicles the insurance companies while calculating the premium to be paid, ascertain the age of the driver, gender, experience and roads the vehicle will be driven for risk assessment. The human element as cause of the accident is always considered prime in such calculations when deciding liability for any accident including those due to drunken driving and over-speeding. To implement AVs, however, the laws must be changed to assess culpability in situations where there is no driver present or when a driver is present but depending on automated controls. China policymakers are looking at shifting liability for AVs being driven in self-driving mode, from driver to manufacturer.

6.5   Summary of AV specific technology related regulations

Under the aegis of the United Nations Economic Commission for Europe (UNECE), around 50 plus nations such as South Korea, Japan and around 42 EU countries have agreed to a common regulation for AVs known as ALKS[825]. This was approved in June 2020[826]. As per UNECE, this law is going to be the first international agreement that is binding on Level 3 AVs and above (driverless) thereby helping the world achieve safe and sustainable mobility for all.

Some of the salient features of ALKS are as follows[827]:

a.   The requirement is to have a Blackbox in all these vehicles which can record every minute detailed recording of the AV's journey and the way the AV was handled during the journey or an accident.

b.   The speed limit of 60 kilometres per hour or 37 miles per hour for all AVs[828].

c.   ALKS to be activated only when the safety driver is behind the steering and has a seatbelt on.

d.   ALKS to be activated only on roads which has a divider and where cycling and walking are forbidden.

e.   As soon as the driver takes control of the vehicle, screens in full AV mode that are monitoring other activities need to be turned off.

---

[825] Countries agree regulations for automated driving Tech Xplore - Technology and Engineering news, https://techxplore.com/news/2020-06-countries-automated.html (last visited April 2, 2021)

[826] Assets.publishing.service.gov.uk, https://assets.publishing.service.gov.uk/government/uploads/system/uploads/attachment_data/file/1044953/prime-minister-to-lord-geidt-21-december.pdf (last visited December 22, 2021).

[827] Compliance with new Automatic Lane Keeping System Regulation, ALKS, Supra Note 498, at 136.

[828] Id.

f. Additionally, the rule requires AV producers to implement Driver Availability Recognition Systems. This will make it easier to keep an eye on the safety driver's ability to regain control of the automobile. The control could be taken over by the driver in many ways which could include methods like spotting eye blinking and closure.

g. ALKS is required to abide by laws governing software updates and cyber-security.

Advanced driver-assistance systems (ADAS) utilise cutting-edge automotive technologies, such as safety features, to assist drivers. ADAS could also use sensors in a vehicle like radars and cameras and then analyse the collected information and provide it to the driver if there is a safety driver in the AV else take automated actions as per the inbuilt software program[829]. If the vehicle is drifting from its lane, ADAS will come to the AV's rescue. The adaptive cruise control, which aids a vehicle in maintaining a safe distance from the car in front of it, is another distinctive element of ADAS. Vehicle-related injuries and fatalities will be significantly decreased with the aid of ADAS. The Autonomous Vehicles Readiness Index (AVRI) is a global tool which is used to evaluate the preparedness of AVs in over 30 countries. AVRI is a comprehensive index which measures 28 individual scores from a variety of sources to translate them into a single score. In most countries, AVRI is generally used by public sector organizations which are responsible for transport and infrastructure. Other public and private sector businesses that deal with or use the road transportation industry also use AVRI.

6.6   Background of Autonomous Vehicles in India

Even though India has adequate knowledge of self-driving AVs, India is completely lagging in the adoption or initiation of steps to adopt AV technology. India has not even commenced the baby steps that many Asian nations have commenced. One of the main reasons for this is the potential job loss for around 2.2 million commercial drivers. Mahindra and Mahindra (M&M) automobiles of India exhibited their indigenously manufactured

---

[829]. 2020 Mobility Insider October 02, What Is ADAS? Aptiv, https://www.aptiv.com/en/insights/article/what-is-adas (last visited April 28, 2021).

autonomous tractor[830]. It is quite evident that to make a self-driving AV that can run in the country M&M will need millions of kilometres of road data and develop an algorithm that will work in Indian conditions. As of date apart from the above announcement by M&M no other automobile Original Equipment Manufacturer (OEM) in India has made any announcement about AV for India.

Considering the topography of India and its population density, it is a no-brainer that to implement an AV ecosystem within India, there will be a great amount of investment required. India has been working on EVs for quite some time to tackle pollution problems. There are many states in India which ply EV buses as public transport. Since India has been the innovation hub for Silicon Valley, there is no doubt that with the background of EV innovation and IT innovation, India can be a hub for AV innovation and export as most nations are looking to integrate Information Technology (IT) capabilities and EV into AV.

In 2020 KPMG released a report depicting the readiness index for AVs also known as Autonomous Vehicle Readiness Index[831] (AVRI). This report analysed around thirty countries on four aspects namely policy and laws, technology and innovation, infrastructure, and customer acceptance. India's position in this report was very bad. It stood last in customer acceptance and AV regulations and just above Brazil in infrastructure availability. As far as road quality goes, India was in the bottom ten lists along with Mexico and the bottom five for innovation. The report also mentioned that the current focus in India is to develop the use of EVs before supporting AV development. The KPMG report also mentioned that the initial usage of AVs and EVs could mainly be in warehouses, ports, and farms. It is said that usage of AVs in farms can reduce dependency on migrant workers which we saw as a key factor during the pandemic. Also, the adoption of AVs and EVs could influence the Insurance and IT sector.

---

[830]. Moneycontrol. 2021. *Autonomous Driving Technology | Can Self-Driving Cars Run on Indian Roads | Moneycontrol.* [online] Available at: <https://www.moneycontrol.com/news/driverless-cars/> [Last visited September 17, 2021].

[831]. 2020. *Autonomous Vehicle Readiness Index.* [ebook] Available at: <https://assets.kpmg/content/dam/kpmg/uk/pdf/2020/07/2020-autonomous-vehicles-readiness-index.pdf> [Last visited September 17, 2021].

India is completely aware of the following factors/data concerning AVs:

a. More than 80% of automobile crashes are caused by human error.

b. Self-driving AVs will help reduce human error and thereby help reduce accidents.

c. Driverless AVs can help reduce fuel/power consumption through vehicle-to-vehicle (V2V) communication and using intelligence on gathered data to adopt a route with less traffic.

d. AVs can help specially challenged persons travel anywhere without depending on other humans.

e. The present trend is to blend AV in EV to help support UNSDG.

f. AVs can also help provide safe travel for women and ensure their safety,

g. Driverless AVs can also be used as delivery vans or for pickup. This will help conserve manpower and time.

6.7   Legal Statutes governing Motor Vehicles in India

The main statute governing motor vehicles in India is the Motor Vehicle Act 1988 and its 2019 amendment along with the Central Motor Vehicle Rules 1989 and various State Motor Vehicle Rules. The MV Act contains legal provisions concerning driver's license, conductor's license, vehicle registration, vehicle permits, traffic rules, vehicle insurance, driver/owner liability, offences and penalties. The MV Act has various provisions to claim compensation for any death or injury caused to a person on account of any accident. As per the MV Act, every vehicle plying on Indian roads must be Insured. Section 140 of, the MV Act introduces a term called "No Fault Liability". As per this section even if an accident or injury or loss or death/disability is caused without the driver/vehicle owner's fault, the vehicle owner must provide compensation. Quite a few amendments were made in 2019 to bring the MV Act on par with global standards. Some of these amendments, which can also be called stepping stones for future amendments to facilitate self-driving AV operations in India are discussed in detail in Chapter 5.

6.8   Suggestions for draft AV Law in India

While the existing MV Act (Act) and MV Rules (Rules) are good enough to support AV operations and testing in India, a few changes are recommended to be made in the Act and Rules for ensuring smooth operation and testing of AVs:

(a) Definition of Driver under Section 2(9) of the Act to be modified to include "Remote Drivers" for self-driving AVs.

(b) Self-Driving AVs are presently not defined under MV Act, and they will have to be defined as a specific class under Section 2(10) of the MV Act.

(c) Section 2(28) of the MV Act covers AVs running on fossil fuel but not on battery. Hence the need to amend this section to include battery-operated self-driving AVs.

(d) Section 7 of the Act deals with "Restrictions on the granting of learner's licences for certain vehicles". As of date, no specific mention of self-driving AVs exists in this section. Once self-driving AVs are defined under Section 2(10), this section will have to be tweaked to include restrictions on granting of learner's licences for self-driving AVs.

(e) Section 10 of the Act deals with "Form and contents of licences to drive". This section lists the type of licences which can be endorsed as a learner license or driving license. Once Section 2(10) is modified to include self-driving AVs then that list must also be included in Section 10.

(f) Section 14 of the Act deals with the "Currency of licences to drive motor vehicles" and this includes a validity period. Should the government decide on a different renewal process/age for self-driving AV driver's licenses then this section will need to be amended.

(g) Section 41 of the Act deals with the registration of vehicles and the period of validity of registration. Should the government decide to reduce or increase the validity period of self-driving AV, then the same should accordingly be annotated in this section through an amendment.

(h) Chapter XI of the Act covers Insurance of Motor Vehicles against Third Party Risks. Based on the development of law in countries where AVs are plying or are being tested one can clearly understand that the law requires manufacturers also to have in place similar insurance during the testing and commercial sale phase of the vehicle as liability for AV accidents can also be attributed to a manufacturing defect. The manufacturer is then required to have back-to-back indemnity with their other vendors and software suppliers. Similar provisions and best practices must be adopted in the India MV Act as and when self-driving AVs get permission for testing and plying in India.

(i) The First Schedule of the Act has mandatory signs under MV Act. A sign needs to be added in this to indicate Entry Prohibition for Self-Driving AVs as there will be in future many crowded areas in India where possibly self-driving AVs may not be permitted. That board will come in handy at that time.

(j) Rule 8A of Rules specifies the minimum training needed to operate an electric rickshaw or electric cart and Rule 9 of Rules prescribes Educational Qualifications for the drivers of goods carriages transporting hazardous or dangerous products. In a similar fashion training and educational qualification requirements, as approved by the government, must be prescribed for drivers of self-driving AV cars whether they are seated in the car or controlling it remotely.

(k) Rule 11 of Rules pertains to Learner's license preliminary test and Rule 15 pertains to the driving test. Both these rules have laid down the syllabus for testing before license grants for conventional motor vehicles. These rules must be changed to include a syllabus for self-driving AVs as and when these AVs are introduced in India.

(l) Rule 123 to Rule 127 pertains to Safety Devices for Drivers, Passengers and Road Users. It would be prudent to mandate through an amendment the installation of a black box or recording facility in self-driving AVs.

Apart from the above-recommended changes, a few changes are also required to allied laws which relate to motor vehicles and AVs once they start plying. Some of the recommended changes are:

(a) National Highways Act 1956, National Highways Rules 1957 and National Highways Authority of India Act 1988 need to be amended to include a definition of AVs and Truck Platooning.

(b) The National Highways Fee (Determination of Rates and Collection) Rules 2008 needs to be amended to cater for the following concerning FASTtag for AVs:

i. Mandatory FASTtag rules for AVs,

ii. Penalty and process to charge a penalty if AVs without FASTtag enter a FASTtag lane or any lane if all lanes are FASTtag lanes,

iii. FASTtag exemption rules for AVs, if being considered by Government,

iv. Connecting bank account / digital payment account of AV owners to FASTtag cards of AV to avoid any payment breach issues,

v. The process to withhold/tow AVs in case of an issue in payment of FASTtag to avoid traffic jams due to lane blocking.

(c) The Control of National Highways (Land and Traffic) Act 2002 must empower the Highway Administration to specifically come out with a different set of guidelines for the removal of unauthorizedly parked driver-driven and driverless vehicles on highways after assessing safety requirements.

(d) Rule 21 of National Highway Administration Rules 2004 needs to be amended to prescribe the process to seize unauthorizedly parked vehicles on highways for any violation including demobilizing them, else they could be remotely controlled and can create havoc in the area where they are seized and parked especially if the owners are anti-nationals/terrorists.

(e) The Central Road Infrastructure Fund Act 2000 provides power to the government to charge excise/cess on petrol/diesel vehicles. This needs to be amended to provide powers to the government to pass orders for charging excise /cess on AVs too. This will make the act future-proof, as normally excise reliefs are provided when introducing new technology and later once the technology is properly accepted into the system, the tax regime follows.

(f) The Central Road Fund (State Funds) Rules 2014 empowers Central Government for utilising the Central Road Fund for specific initiatives, plans, and projects about the construction and upkeep of public roadways. Rule 6 of these rules needs to be amended to include the utilisation of funds for plying of AVs.

(g) The Public Liability Insurance Act and Rules of 1991 presently provide public liability insurance with an intent to provide quick and fast relief to accident victims who are affected when handling hazardous substances and for connected/incidental matters. This Act and Rules need to be amended if AVs are going to be allowed in future to carry hazardous substances.

(h) The Motor Transport Workers Act, of 1961 provides for the welfare of Motor Transport workers and helps regulate their working conditions. This act will need to be amended to include the working conditions of AV drivers.

(i) The Multimodal Transportation of Commodities Act, of 1993 regulates the multimodal transportation of goods through a multimodal transport contract from any site in India to another location outside of India. This law has been drafted, including liability basis, keeping in mind conventional human driver-driven motor vehicles. As per Section 13 of this Act, there shall be no liability on the multimodal transport operator if the operator proves that there was no fault or neglect on his part or part of servants or agents employed by him. Once we have AVs plying albeit in a small portion of the multimodal transportation area say e.g., in ports for transporting from trucks to ships and these AVs end up in an accident, then going by present laws the AV owner could be made liable. Since AV accidents are complicated and involve many players, this act needs to be suitably amended once AVs are introduced in India and are likely to be used in any portion of multimodal transportation.

(j) The Road Transport Corporations Act, of 1950 has been provided for the establishment of and control over Road Transport Corporations. Currently, the definition of road transport service in the act is defined as a service to move products or passengers by road in vehicles for hire or reward. As per the present definition of vehicle and AV can also be said to be included in the definition. The Act still needs to be amended to clearly define AV and elaborate as to which all AVs can be used as Road Transports viz Type 4 with driver or Type 5, driverless, the routes where these AVs can be permitted – long or short or in specific locations like airports, railway stations and ports.

6.9 Product Liability – India and the rest of the world

Product Liability Law, which is more popular in advanced countries, is a law providing legal recourse to consumers for defective products. During the 18th to 20th-century consumers reached out directly to manufacturers for defective products and no court or law was involved. Post 20th century beginning the production style changed, the supply chain evolved, international trade grew, and e-commerce started, and all these led to the development of product liability laws[832] to protect consumers from unethical and fraudulent practices.

Product Liability for self-driving AVs is an evolving area as both the product and the technology are still under development and hence cannot be compared with other existing off-the-shelf products including conventional driver-driven automobiles. Before Consumer Protection Act 2019 or CPA 2019, India had no specific product liability clause in any Indian statute. In India, contract law, applicable portions of the Consumer Protection Act of 1986, the Sale of Goods Act of 1930, the Indian Penal Code of 1860 (presently Bharatiya Nyaya Sanhita), and various other statutes dealing with particular goods and standardisation were the primary legal frameworks for product liability[833]. The CPA 2019 lists out specific responsibilities of the product manufacturer/service provider/seller specifically to compensate consumers for harm caused by a defective product. The word "Consumer" is defined in section 2(7) which includes online consumers and the word "Product" has been defined in section 2(33) of CPA 2019[834].

Under CPA 2019 the term "product" sometimes refers to raw materials or intermediate goods but excludes human body parts. Product liability under CPA 2019 is essentially described as the seller's or manufacturer's obligation to make up for any harm caused to the customer due to the usage of a defective product, especially regarding the quality and safety of the product. Under Section 83 of the CPA 2019, any complaint may bring a product liability lawsuit for a defective product against a manufacturer, a product service provider, or a product vendor. According to Section 84 of CPA 2019, a product maker is only liable in a product liability case if the product has a manufacturing flaw, design defect, or divergence from manufacturing requirements, or the product is in non-conformity with an express warranty or has inadequate instructions concerning its usage. Being non-negligent in making an express warranty of the product is no defence for a manufacturer.

---

832. Don Mayer ET AL., Basics of Product Liability, Sales and Contracts (2012).
833. Anindya Ghosh, Product liability law in India: An evolution - consumer protection - india Welcome to Mondaq (2020), https://www.mondaq.com/india/dodd-frank-consumer-protection-act/974270/product-liability-law-in-india-an-evolution (last visited September 1, 2021).
834. Id.

As per Section 85 of CPA 2019, product service providers shall be liable if the service provided was substandard, flawed, inadequate, or insufficient in quality, nature, or mode of execution as required by or under any law or contract that is now in force. Additionally, a product service provider is responsible for any negligence, carelessness, or intentional withholding of information which resulted in harm or for not issuing adequate warnings/instructions to prevent any harm. A product service provider is also liable if the service did not adhere to the express warranty or the contract's terms and conditions. According to Section 86 of the CPA 2019, a product seller may be held liable for damages if they had significant control over a product's development, testing, production, packaging, or labelling but were otherwise not a product maker. Sellers of goods may also be held accountable if they changed or modified the item and those changes or modifications were the primary factors in the harm. In addition to any explicit warranties provided by the manufacturer, a product seller may also be held accountable if the product fails to live up to the express warranty provided and caused the damage.

Section 87 lays down the exceptions for actions against the product seller and manufacturer for product liability. As per section 87, the product vendor is not responsible if the product was misused, modified, or altered at the time of harm.[835]. As per sub-section (3) of Section 87 CPA 2019:

"Product manufacturer shall not be liable for failure to instruct or warn about a danger which is obvious or commonly known to the user or consumer of such product or which, such user or consumer, ought to have known, taking into account the characteristics of such product"[836] (Section 87 (3) CPA 2019).

6.10 Gaps in Indian Product Liability Law with specific reference to AVs

For any AV crash, one can have two kinds of litigation, a no-fault insurance claim or a product liability claim. The shift in responsibility from the driver

---

835. Bindu Janardhanan Sidharth Sethi, Product liability under the Consumer Protection Act, 2019: Let the manufacturer/seller beware! Bar and Bench - Indian Legal news (2020), https://www.barandbench.com/columns/product-liability-under-the-consumer-protection-act-2019-let-the-manufacturer-seller-beware (last visited September 1, 2021).
836. Id.

to the manufacturer in the case of AVs has made no-fault insurance more attractive than a product liability suit which is more expensive and time-consuming. Globally product liability law is generally a hybrid of contract and tort law making the manufacturer liable for their products. The manufacturers are expected to have back-to-back liability clauses with all their vendors and in the case of AVs also same will be applicable. Wherever we talk of "Strict Product Liability", it would mean that there is no need to prove negligence[837]. Technically there are three kinds of defects viz manufacturing defect, design defect and warning defects (failure to warn)[838].

In the Indian context CPA, 2019 would govern the aspect of Product Liability. As discussed earlier, criminal liability and damages associated with criminal liability and product liability are two different areas. Product liability would be something which the seller/manufacturer is liable to pay up and in case they have multiple vendors to manufacture a product then they must have back-to-back liability contracts with these vendors to take care of any eventuality wherein they have to pay up damages on account of product liability. In the case of AVs, the manufacturer needs to have such back-to-back arrangements with all vendors including but not limited to vendors for various vehicle parts, sensor vendors and software vendors. Criminal liability would be governed by individual culpability and therein a case could be filed separately against the manufacturer, sensor manufacturer, radar manufacturer, and software vendors in the case of an AV.

As of date product liability laws are only in one category and there are no distinct subcategories to take care of self-driving AVs. Under the current legal provisions claims involving manufacturing defects would require that the proof of failure or manufacturing defect be proved, and a correlation established between the defect and the cause of the accident of the self-driving system. When dealing with product liability issues of self-driving AVs one must fully understand various components like Adar sensors, laser rangefinders, video cameras, global positioning, or digital mapping, in an AV and how they interface with each other[839]. If it finally comes out that the error is only a

---

[837] John Villasenor, Products liability and driverless cars: Issues and guiding principles for legislation Brookings (2018), https://www.brookings.edu/research/products-liability-and-driverless-cars-issues-and-guiding-principles-for-legislation/ (last visited October 1, 2021).

[838] James M. Anderson, Autonomous Vehicle Technology: A guide for policymakers (2016).

[839] Cohen IADC product liability May 2015 - pbnlaw.com, (2015), https://pbnlaw.com/media/540169/Cohen-IADC-Product-Liability-May-2015.pdf (last visited October 1, 2021).

software error then technically under the present interpretation, the software may not come under a manufactured product in most product liability laws and hence product liability cannot be claimed in most countries.

Because of the foregoing, CPA 2019 is as on date adequate when it comes to handling product liability cases that might arise due to the usage of self-driving AVs. To handle criminal liability concerning AV accidents, there are tweaks required to be made to the MV Act as discussed earlier.

6.11 Gaps in the Indian Evidence Act and Information Technology Act which can hinder operations of self-driving AVs in India

As per Section 66(1) Information Technology Act 2000, hacking with Computer System is defined as:

> "Whoever with the intent of causing or knowing that is likely to cause wrongful loss or damage to the public or any person destroys or deletes or alters any information residing in a computer resource or diminishes its value or utility or affects it injuriously by any means, commits hacking"[840] (Section 66 (1) IT Act 2000).

However, it can be successfully argued that the hacking of AVs will not count as hacking under the IT Act. This is because the definition of "computer resource" in the IT Act 2000, does not include AVs. This is a clear gap and will have to be plugged as and when self-driving AVs are introduced in India for testing and plying. Apart from correcting the IT Act 2000, it is also required that the definition of hacking in the IT Act 2000 be widened, and strict provisions be incorporated in the said Act to ensure that car manufacturers include effective safety measures and anti-hacking defences in AVs[841]. The IT Act 2000 must also be amended to include the issue of liability and insurance on account of accidents caused due to hacking. The Indian Evidence Act which now is revised as Bharatiya Sakshya Adhiniyam, of date appears to be comprehensive enough to handle any case connected with AVs.

---

[840] The Information Technology Act 2000, Sec 66(1).

[841] The RMLNLU Law Review Blog, Moving towards a driverless era The RMLNLU Law Review Blog (2021), https://rmlnlulawreview.com/2017/10/25/moving-towards-a-driverless-era/ (last visited October 1, 2021).

6.12 ISO 21434:2021

ISO has recently published ISO 21434:2021 standards on cybersecurity engineering on-road vehicles. In terms of cybersecurity risk management, ISO 21434 outlines several specifications, including, but not restricted to, those for the design, creation, manufacture, use, maintenance, and decommissioning of electrical and electronic systems (E / E) in road vehicles, including their parts and interfaces. ISO 21434 in the long run will help manufacturers and their supply chains to be ahead of the curve in protecting their components and road users from cyber-attacks[842].

6.13 Ethical Issues for consideration while programming AVs

Presently in manual-driven vehicles decision on what to avoid or what to save is decided by drivers and these decisions are based on the culture of that country/state and may not be uniform across the world. When it comes to AVs these decisions will be based on how AVs are programmed. Human bias or human ethics based on the region where these vehicles are manufactured can be one strong factor in deciding how these vehicles are to be programmed. But this may not work, or the vehicles may not be popular across the world unless they are programmed separately basis the culture and laws/judicial precedents of the places where these vehicles are to be sold or operated.

6.14 Conclusion

The analysis of the existing legal regime in the USA and developed nations of the EU concerning AV indicates that all these countries have a robust and well-planned scheme to permit plying of AVs in their countries. These countries have very clearly defined who is an Autonomous Vehicle driver, what is an AV, how much insurance is mandatory before testing an AV and how much when plying an AV. Since truck platooning is more popular in the USA, most states in the USA have specific laws for truck platooning and they have exempted the close distance rule for succeeding vehicles in the case of truck platooning this sounds logical as all vehicles in a platoon are controlled by the leading

---

[842.] Rajeev Panicker, ISO 21434:2021 - an overview of newly published standard on Cybersecurity Engineering for Road Vehicles LinkedIn (2021), https://www.linkedin.com/pulse/iso-214342021-overview-newly-published-standard-road-panicker/ (last visited October 1, 2021).

vehicle. Also, since the number of AV accidents has been high in the USA, number-wise, they have been learning from past accidents and accordingly most states are continuously tweaking their laws, especially concerning the safety of vehicles and pedestrians.

Mostly every member state of the Council of Europe has its legal regime for regulating AVs and semi-AVs. Countries like Germany and UK have enacted AV-specific laws, while others have been able to use existing motor vehicle, insurance and liability laws for AVs and Semi AVs. Nevertheless, there are a few pertinent international and regional instruments. Several laws already exist at various levels of government regarding the requirements for vehicle safety. These laws cover a variety of topics, including insurance, product liability, international vehicle standards, criminal and civil penalties for traffic infractions, roadworthiness standards and procedures, consumer information and marketing standards, driving licences, accident investigation processes, data protection laws, regulation of the taxi and private hire markets, regulation of vehicle maintenance, regulation of vehicle repair, regulation of vehicle repair costs, regulation of vehicle maintenance costs, regulation of vehicle maintenance costs, regulation of vehicle maintenance costs (rules of the road)[843].

Legal Liability is another major issue to be addressed for the smooth use of AVs. Accidents caused by AVs are complex to dissect when it comes to liability and different jurisdictions across the world have afforded different treatment to this complex aspect of AV accident liability. The human element as the cause of the accident is always considered prime in such calculations when deciding liability for any accident including those due to drunken driving and over-speeding. However, to use AVs, the laws must be changed to assess liability in situations where there is no driver present or when a driver is present but depending on automated controls. There are countries which are looking at shifting liability for AVs being driven in self-driving mode, from driver to manufacturer. The importance of data generated by AVs is well known and most jurisdictions have laws to protect the same. The requirement of recording devices and the process for handling and storage of data including deletion is spelt out in many laws. Some of the notable ones are Alabama, Utah,

---

[843.] Committee on Legal Affairs and Human Rights Legal aspects of autonomous vehicles, https://assembly.coe.int/LifeRay/JUR/Pdf/DocsAndDecs/2020/AS-JUR-2020-20-EN.pdf (Last visited on April 12, 2021)

Germany, and Russia. As per Russian regulations, every AV must mandatorily have a data recording system and the driver must also be able to switch off and on the equipment for traffic recording and actions of the driver.

It can be summarized that the technology for semi-AV and full AV is well-developed, and the stage is set for commercial deployment as AVs are going to be a part and parcel of human life soon. However, each country has budgetary, policy and regulatory issues which must be solved so that AV deployments can happen in full steam. The speed of AV deployment in a country would depend on the way the Government regulates AVs in that country. Governments must support road trials through well-defined regulatory priorities and proper control of AV policies, and this will help speed up AV deployment. Product liability and regulations for self-driving AVs is likely to evolve gradually over a period. To have self-driving AVs ply in India, there are a lot of legal and infrastructural hurdles that will have to be navigated[844]. In the Indian context, while self-driving AVs can be a dystopian nightmare, they will help us experience a better quality of driving, ensure safe driving, and reduce traffic congestion and driving stress. However, the fear of sensors failing especially when driving in congested areas in India can be stressful. Product Liability which has been introduced in CPA 2019 is expected to mature in the years to come and one will see a lot of judgements in that area which will act as precedents going forward. Lastly, adequate data protection and privacy safeguards must also be provided to all users of AVs, if it is not covered under general data protection laws[845].

* * *

---

[844] Autonomous Driving Technology: Can self-driving cars run on Indian Roads, Moneycontrol (2020), https://www.moneycontrol.com/news/driverless-cars/ (last visited October 1, 2021).

[845] Diganth Raj Sehgal - et al., Legal issues related to Autonomous Vehicles iPleaders (2020), https://blog.ipleaders.in/legal-issues-related-autonomous-vehicles/ (last visited October 17, 2021).

BIBLIOGRAPHY

1. Primary Sources
A. Cases

Airbus Industrie v. Laura Howell Linton (1994) - Court agreed to handle the product liability case even though we had no law on that in India at that time, ILR 1994 KAR 1370.

A. K. Roy & Another vs Voltas Limited – AC making noise, once replaced no more remedy, 1973 AIR 225.

A.S. Mittal v. State of U.P. (1989) – Mass Eye Surgery case – Was a pre-test of eye saline done? 1989 AIR 1570.

Atul Anant Rane & Ors. Vs. Lourdes Gas Service & Ors – Death due to wrong handling of the Gas knob – No Remedy, Complaint Nos. 34 & 288 of 1993, decided on, 04 November 1997 at, Maharashtra State Consumer Disputes Redressal Commission Mumbai.

EID (Parry) India Ltd. v. Baby Benjamin Tushara – Damage to commode after 30 years of use, 1992 (23) DRJ 394.

Escola v. Coca Cola Bottling Co. (1944) – Manufacturer responsible for Defective Product, 24 Cal.2d 453, 150 P.2d 436 (Cal. 1944).

G. Govindan vs New India Assurance Co. Ltd. on 8 April 1999, 1999 (2) SCR 476.

Greenman v. Yuba Power Products, Inc and Henningsen v. Bloomfield Motors, Inc – Manufacturer responsible for defect/damage if the product was used as per specification, 59 Cal.2d 57 (Cal. 1963) and 32 N.J. 358 (N.J. 1960).

Justice K.S. Puttaswamy (Retd) vs Union of India and Ors. on 24 August 2017), AIR 2017 SC 4161.

K. Gopalakrishnan Minor vs Sankara Narayanan and Ors. on 4 October, 1967, AIR 1968 Mad 436.

Kharak Singh vs The State of U. P. & Others 1963 AIR 1295, 1964 SCR (1) 332

MacPherson v. Buick Motor Co (1916) – Wooden Wheel Collapsed – No Privity of Contract ruled, 111 N.E.1050 (1916).

Madineni Kondaiah & Ors. v. Yaseen Fatima & Ors. on June 22, 1985, 1 (1986) ACC 501

MC Mehta & Another versus Union of India in the Bhopal gas case,1987 AIR 1086

M. P. Sharma and Others vs Satish Chandra and Others 1954 AIR 300, 1954 SCR 1077

M/s. Solidaire India Ltd. v/s D.A. Mohan Kumar – TV Picture tube damaged after the warranty period. No remedy but the company gave a discount for a new tube - Appeal No. 18 of 1990, decided on, 29 June 1991 at, Kerala State Consumer Disputes Redressal Commission Thiruvananthapuram.

National Insurance Co. Ltd. vs Deorao and Ors. on 8 July, 2002, II (2004) ACC 328.

National Insurance Co. Ltd. vs Faqir Chand and Ors. on 25 April, 1994, AIR 1995 J K 91

Oriental Insurance Co. Ltd vs Inderjit Kaur & Ors on 8 December 1997, 1998 SCC (1) 371.

Puttaswamy vs UOI 2017 - (2017) 10 SCC 1

People's Union for Civil Liberties (PUCL) v. Union of India, (1997) 1 SCC 301.

R v Gold & Schifreen [1988] 1 AC 1063 (HL), [1988] 2 All ER 186, [1988] 2 WLR 984,

S. Sudhakaran vs A.K. Francis and Ors. on 12 June, 1996, AIR 1997 Ker 26.

Sohan Lal Passi vs P. Sesh Reddy & Ors on 17 July 1996, 1996 SCC (5) 21.

Winterbottom v. Wright (1842) – Contract for Mail Coach Service Privity of Contract Case, (1842) 10 M&W 109).

b. Statutes

Australia:

Electronic Transactions Act 1999

Costa Rica:

Digital Signature Law 2005

India:

Central Motor Vehicle Rules, 1989

Central Road Fund (State Roads) Rules, 2014

Central Road and Infrastructure Fund Act, 2000

Constitution of India 1950

Competition Act, 2002, No 12 of 2003

Consumer Protection Act, 2019, No 35 of 2019

Control of National Highways (Land and Traffic) Act, 2002

Credit Information Companies (Regulation) Act, 2005

Criminal Procedure Code 1973, Act No 2 of 1974 (presently Bharatiya Nagarik Suraksha Sanhita)

Driving License (Conditions for Exemption) Rules, 1992

Fatal Accidents Act, 1855

Indian Evidence Act 1872 (presently Bharatiya Sakshya Adhiniyam)

Information Technology Act 2000,

Information Technology (Amendment) Act 2008,

Information Technology (Reasonable security practices and procedures and sensitive personal data or information) Rules 2011

Liquefied Petroleum Gas (Regulation of Use in Motor Vehicles) Order, 2001

Maharashtra Control of Organised Crimes Act 1999,

Motor Vehicles (Amendment) Act 2019, Act No 32 of 2019

Motor Vehicles Act 1988, Act No 59 of 1988

Motor Vehicles (All India Permit for Tourist Transport Operators) Rules, 1993

Motor Vehicles (Driving) Regulations, 2017

Motor Vehicles (High Security Registration Plates) order, 2018

Motor Vehicles (Vehicle Location Tracking Device and Emergency Button) Order, 2018

Motor Transport Workers Act, 1961

Multimodal Transportation of Goods Act, 1993

National Highways Act, 1956

National Highways Authority of India Act, 1988
National Highways Administration Rules, 2004
National Highways Rules, 1957
Overall Dimensions of Motor Vehicles (Prescription of Conditions for Exemption) Rules, 1991
Police and Criminal Evidence Act 1984
Prevention of Terrorism Act 2002,
Public Liability Insurance Act, 1991
Public Liability Insurance Rules, 1991
Road Transport Corporations Act, 1950
Rent a Cab Scheme, 1989
Rules of the Road Regulations, 1989
Solatium Scheme, 1989

Malaysia:
Malaysia Computer Crimes Act
Malaysia Digital Signature Act

Mexico:
E-Commerce Act 2000

Thailand:
Thailand Computer Crimes Act

UK:
Computer Misuse Act 1990
Electronic Communications Act 2000

USA:
California Consumer Privacy Act 2018
California Code of Regulations Title 13
Driver's Privacy Protection Act 1994
Electronic Communication Privacy Act 1986
Foreign Intelligence Surveillance Act 1978
Gramm-Leach-Bliley Act 1999
Homeland Security Act 2002
Intelligence Reform and Terrorism Prevention Act 2004
Nevada Assembly Bill No. 511, March 28, 2011

Pen Register Act 1986
Privacy Act 1974
Stored Communications Act 1986
Wiretap Act 1968

c. Legislation (UK, US, India)
European Union:
Electronic Signatures Directive (1999/93/EC)

d. Foreign Legislatures
(Federal Automated Vehicles Policy2016) < https://www.transportation.gov/sites/dot.gov/files/docs/AV policy guidance PDF.pdf> last visited January 20, 2020

e. Bills
Personal Data Protection Bill, India 2019

2. Secondary Sources
a. Books
Anderson JM and others, Autonomous Vehicle Technology: A Guide for Policymakers (Rand Corporation 2016)
Barlow JP, A Declaration of the Independent Cyberspace
Constantine S, Report on the Attribution of Civil Liability for Accidents Involving Autonomous Cars (Law Reform Committee, Singapore Academy of Law 2020)
Daniel C Vladeck, Machines Without Principals: Liability Rules and Artificial Intelligence (2014)
Don Mayer et al., Basics of Product Liability, Sales and Contracts (2012).
Duggal Pavan, Artificial Intelligence Law (2017).
Johnson DR and Post D, Law and Borders: The Rise of Law in Cyberspace (2006)
Nick Bostrom, Superintelligence: Paths, Dangers, Strategies (OUP 2014

Raysman R, Emerging Technologies and the Law: Forms and Analysis (Law Journal Press 2002)

Raysman R and Brown P, Computer Law: Drafting and Negotiating Forms and Agreements (Law Journal Press 2003)

Simester AP and others, Criminal Law: Theory and Doctrine (Hart Publishing 2016)

Solove DJ and Schwartz PM, Privacy, Information, and Technology (Wolters Kluwer Law & Business 2018)

Trout BJ, Cyber Law: A Legal Arsenal for Online Business (World Audience 2007)

Weaver JF, Robots Are People Too: How Siri, Google Car, and Artificial Intelligence Will Force Us to Change Our Laws (Praeger 2014)

Zavrsnik A, "Drones and Unmanned Aerial Systems" (Springer International Publishing, AG Switzerland 2016)

b. Journal Articles

Abu Kassim et al., Is Malaysia Ready to Adopt Autonomous Vehicles? 3 Journal of Society of Automotive Engineers Malaysia 84–88 (2019).

Artificial Intelligence for sustainable and effective justice delivery in India Parth Jain Department of Laws, Panjab University, Chandigarh, India Corresponding author: adv.parth.jain@gmail.com, parthjainjudge@gmail.com © Authors et al OIDA International Journal of Sustainable Development, Ontario International Development Agency, Canada ISSN 1923-6654 (print) ISSN 1923-6662 (online) www.oidaijsd.com Also available at http://www.ssrn.com/link/OIDA-Intl-Journal-Sustainable-Dev.html

Jain P, "Artificial Intelligence for Sustainable and Effective Justice Delivery in India" (2018) II OIDA International Journal of Sustainable Development 63 <https://papers.ssrn.com/sol3/papers.cfm?abstract_id=3284903> last visited January 20, 2020

Kemp, Richard, "Legal Aspects of Artificial Intelligence – Kemp IT Law", 2018.

Ayog, Niti, "National Strategy on Artificial Intelligence | NITI Aayog", 2018.

Solaiman, S. M., "Legal personality of robots, corporations, idols and chimpanzees: a quest for legitimacy" (2017) 25 Artificial Intelligence and Law.

T. Vikram, "Cyber Crimes- A Study with a Case", July-September 2002, Indian Police Journal 78.

Warren SD and Brandeis LD, "The Right to Privacy" (1890) 4 Harvard Law Review 193

c. *Online Articles*

Autonomous Vehicles Meet Human Drivers – GHSA, https://www.ghsa.org/sites/default/files/2017-01/AV 2017 - Final.pdf (Last visited on February 2, 2020)

Automated Driving Systems, https://www.nhtsa.gov/sites/nhtsa.dot.gov/files/documents/13069a-ads2.0_090617_v9a_tag.pdf (Last visited on February 2, 2020)

d. *Encyclopedia*

Pickover, C., 2019. Artificial Intelligence: An Illustrated History. Sterling Publishing Co., Inc.

e. *Websites and Blogs*

1998(4) all mr (journal) 20, Atul Anant Rane & Ors. vs. Lourdes Gas Service & Ors., NearLaw.com (1997), https://nearlaw.com/PDF/MumbaiHC/1998/1998(4)-ALL-MR-(JOURNAL)-20.html (Last visited on September 1, 2021).

2017-2018 BILL 3289: Safe following distance, https://www.scstatehouse.gov/sess122_2017-2018/bills/3289.htm (Last visited on December 28, 2020).

2019 Autonomous Vehicles Readiness Index (2019), https://assets.kpmg/content/dam/kpmg/xx/pdf/2019/02/2019-autonomous-vehicles-readiness-index.pdf (Last visited 29 April 2021).

2019. Motor Vehicle Amendment Act 2019. [ebook] Government of India. Available at: <https://egazette.nic.in/WriteReadData/2019/210413.pdf> [Last visited 17 August 2021].

2019. MVA Bill 2019 Analysis. [ebook] SLF. Available at: <https://savelifefoundation.org/wp-content/uploads/2019/07/MVA-Bill-2019_Analysis_SLF.pdf> [Last visited 17 August 2021].

2020 Autonomous Vehicles Readiness Index (2020), https://assets.kpmg/content/dam/kpmg/it/pdf/2020/07/2020-autonomous-vehicles-readiness-index.pdf (Last visited on April 30, 2021).

2020. Autonomous Vehicle Readiness Index. [ebook] Available at: <https://assets.kpmg/content/dam/kpmg/uk/pdf/2020/07/2020-autonomous-vehicles-readiness-index.pdf> [Last visited 17 September 2021].

2020 Mobility Insider October 02, What Is ADAS? Aptiv, https://www.aptiv.com/en/insights/article/what-is-adas (Last visited on April 28, 2021).

6 Ways AI and Robotics Are Improving Healthcare (Robotics Business Review May 29, 2019), https://www.roboticsbusinessreview.com/health-medical/6-ways-ai-and-robotics-are-improving-healthcare (Last visited on February 24, 2020).

7 benefits of Autonomous Cars, Thales Group (2017), https://www.thalesgroup.com/en/markets/digital-identity-and-security/iot/magazine/7-benefits-autonomous-cars (last visited December 12, 2021).

Abhilasha Singh, Internet-connected cars: What data is collected, who has the access and more The Financial Express (2021), https://www.financialexpress.com/auto/car-news/internet-connected-cars-what-data-is-collected-access-hyundai-venue-kia-sonet-seltos-mg-hector/2305516/ (Last visited on Oct 1, 2021).

Autonomous Driving Technology: Can self-driving cars run on Indian Roads, Moneycontrol (2020), https://www.moneycontrol.com/news/driverless-cars/ (Last visited on October 1, 2021).

Accenture Insurance, Around the world in self-driving cars: Asia Accenture Insurance Blog (2017), https://insuranceblog.accenture.com/around-the-world-in-self-driving-cars-asia (Last visited on April 27, 2021).

Ahmad and others, The Americas and the Caribbean (Regulation of Artificial Intelligence January 1, 2019), https://www.loc.gov/law/help/artificial-intelligence/americas.php (Last visited on January 17, 2020).

An Act concerning autonomous vehicles, https://www.cga.ct.gov/2017/ACT/pa/2017PA-00069-R00SB-00260-PA.htm (Last visited on December 24, 2020).

Andrew J. Hawkins, Uber has resumed testing its self-driving cars in San Francisco The Verge (2020), https://www.theverge.com/2020/3/10/21172213/uber-self-driving-car-resume-testing-san-francisco-crash (Last visited on December 30, 2020).

Andrew J. Hawkins, Uber self-driving car saw pedestrian but didn't brake before fatal crash, feds say The Verge (2018), https://www.theverge.com/2018/5/24/17388696/uber-self-driving-crash-ntsb-report (Last visited on December 30, 2020).

Anindya Ghosh, Product liability law in India: An evolution - consumer protection - india Welcome to Mondaq (2020), https://www.mondaq.com/india/dodd-frank-consumer-protection-act/974270/product-liability-law-in-india-an-evolution (Last visited on September 1, 2021).

Anne Teigen Ben Husch, Regulating Autonomous Vehicles, https://www.ncsl.org/research/transportation/regulating-autonomous-vehicles.aspx (Last visited on January 10, 2021).

Artificial Intelligence and Its Impact on Cyber Security (Infosec Resources May 1, 2017), https://resources.infosecinstitute.com/artificial-intelligence-impact-cyber-security/#gref (Last visited on January 17, 2020)

Artificial Intelligence and Privacy (Artificial Intelligence and Privacy January 2018), https://iapp.org/media/pdf/resource_center/ai-and-privacy.pdf (Last visited on January 17, 2020)

Assets.publishing.service.gov.uk, https://assets.publishing.service.gov.uk/government/uploads/system/uploads/attachment_data/file/1044953/prime-minister-to-lord-geidt-21-december.pdf (last visited December 22, 2021).

Autonomous connected cars on their way, archive.shine.cn (2016), https://archive.shine.cn/business/biz special/Autonomous-connected-cars-on-their-way/shdaily.shtml (Last visited on April 25, 2021).

Autonomous Vehicles, GHSA, https://www.ghsa.org/taxonomy/term/521 (Last visited on January 10, 2021).

Automated vehicles driving and testing SB 995-998, https://www.legislature.mi.gov/documents/2015-2016/billanalysis/Senate/pdf/2015-SFA-0995-N.pdf (Last visited on December 26, 2020).

Autonomous Vehicles, GHSA, https://www.ghsa.org/taxonomy/term/521 (Last visited on January 2, 2021).

Autonomous vehicles in Indonesia, Inside Tech Law | Global law firm | Norton Rose Fulbright, https://www.insidetechlaw.com/autonomous-vehicles/08_indonesia#:~:text=Indonesia%20currently%20does%20not%20have,Informatics%20(MOCI)%20Regulation%20No. (Last visited on April 30, 2021).

Autonomous vehicles law and regulation in China, CMS Expert Guides, https://cms.law/en/int/expert-guides/cms-expert-guide-to-autonomous-vehicles-avs/china (Last visited on April 29, 2021).

Autonomous vehicles law and regulation in Singapore, CMS, https://cms.law/en/int/expert-guides/cms-expert-guide-to-autonomous-vehicles-avs/singapore (Last visited on April 30, 2021).

Automated Vehicle Testing in Vermont, | Agency of Transportation, https://vtrans.vermont.gov/planning/av-testing#:~:text=Vermont's%20Automated%20Vehicle%20Testing%20Act,on%20state%20and%20town%20highways. (Last visited on December 28, 2020).

Autonomous vehicles law and regulation in Turkey CMS Expert Guides, https://cms.law/en/int/expert-guides/cms-expert-guide-to-autonomous-vehicles-avs/turkey (Last visited on March 22, 2021)

Autonomous vehicles law and regulation in Ukraine CMS Expert Guides, https://cms.law/en/int/expert-guides/cms-expert-guide-to-autonomous-vehicles-avs/ukraine (Last visited on April 2, 2021)

Autonomous Vehicle (AV) Policy Framework, Part I: Cataloging Selected National and State Policy Efforts to Drive Safe AV Development, https://innovationisrael.org.il/sites/default/files/Autonomous%20Vehicle%20Policy%20Framework-.pdf (Last visited on March 12, 2021)

Bag A, Presentation of Electronic Evidence in Court in Light of the Supreme Court Judgment in Anvar P. K. vs. P.K Basheer & Ors. (iPleaders September 27, 2014), https://blog.ipleaders.in/presentation-of-electronic-evidence-in-a-court-in-light-of-the-supreme-court-judgment-in-ansar-p-k-vs-p-k-basheer-ors (Last visited on January 17, 2020)

Barua P, Artificial Intelligence and Law (Law Times Journal April 24, 2019), http://lawtimesjournal.in/artificial-intelligence-and-law (Last visited on January 17, 2020)

Betsy Foresman, Utah to join states allowing fully autonomous vehicles on public roads StateScoop (2019), https://statescoop.com/utah-to-join-states-allowing-fully-autonomous-vehicles-on-public-roads/ (Last visited on December 28, 2020).

Bindu Janardhanan Sidharth Sethi, Product liability under the Consumer Protection Act, 2019: Let the manufacturer/seller beware! Bar and Bench - Indian Legal news (2020), https://www.barandbench.com/columns/product-liability-under-the-consumer-protection-act-2019-let-the-manufacturer-seller-beware (Last visited on September 1, 2021).

Bill Covington, Legislating Autonomous Vehicles in Washington (2018), https://wstc.wa.gov/wp-content/uploads/2019/11/2018-0717-BP7-UWFullReportAVLawScan.pdf (Last visited on December 30, 2020).

Bill Resource, https://custom.statenet.com/public/resources.cgi?id=ID%3Abill%3ALA2018000H308&ciq=ncsl&client_md=569a3280675b8f459612254d497b0650&mode=current_text (Last visited on December 26, 2020).

Bill Text - AB-669 Department of Transportation: motor vehicle technology testing, https://leginfo.legislature.ca.gov/faces/billNavClient.xhtml?bill_id=201720180AB669 (Last visited on December 22, 2020).

Bill Text - AB-1444 Livermore Amador Valley Transit Authority: demonstration project, https://leginfo.legislature.ca.gov/faces/billTextClient.xhtml?bill_id=201720180AB1444 (Last visited on December 22, 2020).

CEE Legal Matters. 2021. Autonomous Vehicles Regulation in Russia. [online] Available at: <https://ceelegalmatters.com/cms/13891-autonomous-vehicles-regulation-in-russia> [Last visited 28 June 2021].

Charlotte Trueman, The state of autonomous vehicles in Southeast Asia CIO (2019), https://www.cio.com/article/3309917/the-state-of-autonomous-vehicles-in-southeast-asia.html (Last visited on April 29, 2021).

Cindy M Grimm, William D Smart and Woodrow Hartzog, An Education Theory of Fault for Autonomous Systems, (WeRobot 2017 conference, New Haven, Mars 2017), www.werobot2017.com/wp-content/uploads/2017/03/Smart-Grimm-Hartzog-Education-We-Robot.pdf (Last visited on February 2, 2020).

Cohen IADC product liability May 2015 - pbnlaw.com, (2015), https://pbnlaw.com/media/540169/Cohen-IADC-Product-Liability-May-2015.pdf (Last visited on October 1, 2021).

Colin Wood, Slow and steady, Indiana readies autonomous vehicle framework StateScoop (2018), https://statescoop.com/slow-and-steady-indiana-readies-autonomous-vehicle-framework/ (Last visited on December 24, 2020).

Committee on Legal Affairs and Human Rights Legal aspects of autonomous vehicles, https://assembly.coe.int/liferay/jur/Pdf/DocsAndDecs/2020/AS-JUR-2020-20-EN.pdf (Last visited on April 12, 2021)

Compliance with new Automatic Lane Keeping System Regulation, ALKS, https://www.tuvsud.com/en-in/industries/mobility-and-automotive/automotive-and-oem/autonomous-driving/compliance-with-new-automated-lane-keeping-system-regulation (Last visited on April 12, 2021)

Congress Taking Another Look at Regulating Automated Driving Systems, JD Supra, https://www.jdsupra.com/legalnews/congress-taking-another-look-at-51054/ (Last visited on December 17, 2020).

Consequential Robotics, 8 Ethical and Societal Issues Relevant to the Future Use of Robots in Social Care (Consequential Robotics, March 12, 2018), http://consequentialrobotics.com/blog/2018/3/12/8-ethical-and-societal-issues-relevant-to-the-use-of-robotics-autonomous-systems-in-social-care (Last visited on February 2, 2020)

Countries agree regulations for automated driving Tech Xplore - Technology and Engineering news, https://techxplore.com/news/2020-06-countries-automated.html (Last visited on April 2, 2021)

COVID-19 + Healthcare (Juniper Networks), https://www.juniper.net/us/en/ai-driven-networking-empowers-effective-coronavirus-surge-response/?utm_medium (Last visited on August 6, 2020)

Crane C, Artificial Intelligence in Cyber Security: The Savior or Enemy of Your Business? (Hashed Out by The SSL Store™ July 30, 2019), https://www.thesslstore.com/blog/artificial-intelligence-in-cyber-security-the-savior-or-enemy-of-your-business (Last visited on January 17, 2020)

Cyber Crime Investigation Manual, https://uppolice.gov.in/writereaddata/uploaded-content/Web_Page/28_5_2014_17_4_36_Cyber_Crime_Investigation_Manual.pdf, (Last visited on February 2, 2020)

Diganth Raj Sehgal - et al., Legal issues related to Autonomous Vehicles iPleaders (2020), https://blog.ipleaders.in/legal-issues-related-autonomous-vehicles/ (Last visited on October 17, 2021).

Ec.europa.eu. 2021. [online], https://ec.europa.eu/transport/sites/transport/files/3rd-mobility-pack/com20180283_en.pdf (Last visited on March 02, 2021).

Elanjchezhien, Right to Privacy (Legal Articles in India March 21, 2018), http://www.legalservicesindia.com/law/article/10/28/Right-To-Privacy (Last visited on January 17, 2020)

Emily Ashcraft, Legislature approves bill allowing driverless cars on Utah roads Deseret News (2019), https://www.deseret.com/2019/3/1/20667185/legislature-approves-bill-allowing-driverless-cars-on-utah-roads (Last visited on December 28, 2020).

Fact sheet: Federal automated vehicles policy overview, https://www.transportation.gov/sites/dot.gov/files/docs/DOT_AV_Policy.pdf (Last visited on January 10, 2021).

Federal Automated Vehicles Policy - September 2016: US Department of Transportation (U.S. Department of Transportation), https://www.transportation.gov/AV/federal-automated-vehicles-policy-september-2016 (Last visited on January 17, 2020)

Florida House of Representative CS/HB 311, https://www.flsenate.gov/Session/Bill/2019/311/BillText/er/PDF (Last visited on December 24, 2020).

Florida Senate Committee of Infrastructure and Security CS/HB 311, https://www.flsenate.gov/PublishedContent/Session/2019/BillSummary/Infrastructure_IS0311is_0311.pdf (Last visited on December 24, 2020).

France pushes for 'highly automated' vehicles by 2022 Automotive News Europe, https://europe.autonews.com/article/20180808/ANE/180809840/france-pushes-for-highly-automated-vehicles-by-2022 (Last visited on March 2, 2021)

France to amend legislation for autonomous vehicle trials | Autovista Group, https://autovistagroup.com/news-and-insights/france-amend-legislation-autonomous-vehicle-trials (Last visited on March 2, 2021)

Fung B, Google's Driverless Cars Are Now Legally the Same as a Human Driver (The Washington Post February 10, 2016), https://www.washingtonpost.com/news/the-switch/wp/2016/02/10/googles-driverless-cars-are-now-legally-the-same-as-a-human-driver (Last visited on January 17, 2020)

Garson J, Who Is Responsible When Robots Kill? (Forbes, February 11, 2019), https://www.forbes.com/sites/jackgarson/2019/02/11/who-is-responsible-when-robots-kill/#4917a16d6b82 (Last visited on February 2, 2020)

General Assembly of North Carolina - House Bill No 469, https://www.ncleg.net/Sessions/2017/Bills/House/PDF/H469v7.pdf (Last visited on December 28, 2020).

Government.ru. 2021. *О проведении эксперимента по эксплуатации на автодорогах высокоавтоматизированных транспортных средств.* [online] Available at: <http://government.ru/docs/34831/> [Last visited 28 June 2021].

Green light for Experimental Law for testing self-driving vehicles on public roads News item | Government.nl, https://www.government.nl/latest/news/2019/07/02/green-light-for-experimental-law-for-testing-self-driving-vehicles-on-public-roads (Last visited on March 12, 2021)

Group Discussion Ideas. 2021. Should Driverless cars be allowed in India? ~ Group Discussion Ideas. [online] Available at: <https://www.groupdiscussionideas.com/should-driverless-cars-be-allowed-in-india/> [Last visited 17 August 2021].

History of the Autonomous Car, TitleMax (2020), https://www.titlemax.com/resources/history-of-the-autonomous-car/#:~:text=History of Autonomous Cars, made this concept a reality (Last visited on January 10, 2021).

H.R. 3388 (115th): Self Drive Act, GovTrack.us, https://www.govtrack.us/congress/bills/115/hr3388/summary (Last visited on December 17, 2020).

HT Auto Desk, Indian contingent at Tokyo Olympics to ride in these electric driverless cars Hindustan Times Auto News (2021), https://auto.hindustantimes.com/auto/news/indian-contingent-at-tokyo-olympics-to-ride-in-these-electric-driverless-cars-41626768980716.html (last visited December 22, 2021).

Hungary Decree No 5/1990, Article §2 (3b), point b.

Hungary Decree No. 6/1990, Annex 17.

IndraStra, An Analysis of Puttaswamy: The Supreme Court's Privacy Verdict (Medium November 18, 2017), https://medium.com/indrastra/an-analysis-of-puttaswamy-the-supreme-courts-privacy-verdict-53d97d0b3fc6 (Last visited on January 17, 2020).

ITS 2019 (2019), https://www.kantei.go.jp/jp/singi/it2/kettei/pdf/20190607/siryou9.pdf, p.21 (Last visited on April 30, 2021).

James M Anderson & Nidhi Kalra, Liability Implications of Autonomous Vehicle Technology JSTOR (2014), https://www.jstor.org/stable/10.7249/j.ctt5hhwgz.14?seq=1#metadata_info_tab_contents (Last visited on September 1, 2021).

Jeff Oslon & Huong Phan, Automotive in Vietnam Lexology (2019), https://www.lexology.com/library/detail.aspx?g=94f93be9-e1be-4d62-a44c-8c6995229650 (Last visited on May 5, 2021).

John Villasenor, Products liability and driverless cars: Issues and guiding principles for legislation Brookings (2018), https://www.brookings.edu/research/products-liability-and-driverless-cars-issues-and-guiding-principles-for-legislation/ (Last visited on October 1, 2021).

Josh Mandell @joshuamandell et al., Virginia still a blank slate for self-driving car laws Charlottesville Tomorrow â€¢ Informed citizens create better communities, https://www.cvilletomorrow.org/articles/virginia-blank-slate-for-self-driving-car-laws (Last visited on February 20, 2021).

Key Changes in the Personal Data Protection Bill, 2019 from the Srikrishna Committee Draft (SFLC.in), https://sflc.in/key-changes-personal-data-protection-bill-2019-srikrishna-committee-draft (Last visited on January 17, 2020)

KPMG Autonomous Vehicles Readiness Index, 2019, https://assets.kpmg/content/dam/kpmg/xx/pdf/2019/02/2019-autonomous-vehicles-readiness-index.pdf (Last visited on March 22, 2021)

Legalserviceindia.com. 1999. The Motor Vehicles Act, 1988. An Analysis. [online] Available at: <https://www.legalserviceindia.com/legal/article-2663-the-motor-vehicles-act-1988-an-analysis.html> [Last visited 1 September 2021].

Legalserviceindia.com. 2021. Motor Vehicle Act,2020 (Amended). [online] Available at: <http://www.legalserviceindia.com/legal/article-4188-motor-vehicle-act-2020-amended-.html> [Last visited 18 August 2021].

Legislative Assembly of North Dakota House Bill No 1065, https://www.legis.nd.gov/assembly/64-2015/documents/15-0167-03000.pdf (Last visited on December 28, 2020).

Legislative Assembly of North Dakota House Bill No 1202, https://www.legis.nd.gov/assembly/65-2017/documents/17-0711-04000.pdf (Last visited on December 28, 2020).

Legislative Assembly of Oregon House Bill No 4059, https://olis.leg.state.or.us/liz/2018R1/Downloads/MeasureDocument/HB4059/Enrolled (Last visited on December 28, 2020).

Legislative Data Processing Center, 2016 Act 101 The official website for the Pennsylvania General Assembly., https://www.legis.state.pa.us/cfdocs/legis/li/uconsCheck.cfm?yr=2016&sessInd=0&act=101# (Last visited on December 28, 2020).

Leon H, Artificial Intelligence Is on the Case in the Legal Profession" (Observer October 16, 2019), https://observer.com/2019/10/artificial-intelligence-legal-profession (Last visited on January 17, 2020)

Linkedin.com. 2019. No Fault Liability. [online] Available at: <https://www.linkedin.com/pulse/fault-liability-major-prashant-rai/> [Last visited 18 August 2021].

Mallesons K& W, China Issues Self-Driving Car Road Testing Regulations (China Law Insight July 27, 2020), https://www.chinalawinsight.com/2018/04/articles/compliance/china-issues-self-driving-car-road-testing-regulations (Last visited on October 13, 2020)

Martin Kahl, Autonomous vehicles: Driving regulatory and liability challenges Automotive World (2020), https://www.automotiveworld.com/articles/autonomous-vehicles-driving-regulatory-and-liability-challenges/ (Last visited on September 1, 2021).

Matthews K, Legal Implications of Driverless Cars (American Bar Association December 2018), https://www.americanbar.org/news/abanews/publications/youraba/2018/december-2018/legal-implications-of-driverless-cars (Last visited on October 13, 2020)

Medium. 2021. How Will State Regulations Fill the Gap for Self-Driving Cars. [online] Available at: <https://mapanauta.medium.com/how-will-state-regulations-fill-the-gap-for-self-driving-cars-9030729e2c7f> [Last visited 8 November 2021].

Melanie Musson Published Insurance Expert, Which states allow self-driving cars? (2020 Update) AutoInsurance.org | Compare auto insurance companies, search 100's of auto insurance reviews, and get cheap auto insurance quotes online. Save $ $ $ on your auto insurance rates! (2021), https://www.autoinsurance.org/which-states-allow-automated-vehicles-to-drive-on-the-road/#5 (Last visited on December 17, 2020).

Merger of FCA and Groupe PSA https://www.fcagroup.com/en-US/media_center/fca_press_release/FiatDocuments/2021/January/The_merger_of_FCA_and_Groupe_PSA_has_been_completed.pdf (Last visited on March 2, 2021)

Mojito Fde and Medha, Cases in Which India's Supreme Court Will Define Contours of Free Speech Online (Centre for Communication Governance, National Law University Delhi September 19, 2013), https://ccgdelhi. org/2013/09/19/cases-in-which-indias-supreme-court-will-define-contours-of-free-speech-online (Last visited on January 17, 2020)

Moneycontrol. 2021. Autonomous Driving Technology | Can Self-Driving Cars Run on Indian Roads | Moneycontrol. [online] Available at: <https://www.moneycontrol.com/news/driverless-cars/> [Last visited 17 September 2021].

M/s. Solidaire India Ltd. v D.A. Mohan Kumar on 29 June ..., (1999), https://www.lawyerservices.in/Ms-Solidaire-India-Ltd-Versus-DA-Mohan-Kumar-1991-06-29 (Last visited on September 1, 2021).

Nebraska senator floats bill to address driverless vehicle liability, ROADS & BRIDGES, https://www.roadsbridges.com/nebraska-senator-floats-bill-address-driverless-vehicle-liability (Last visited on February 16, 2021).

Need for a Regulatory Framework for the National AI Marketplace, https://indiaai.in/article/need-for-a-regulatory-framework-for-the-national-ai-marketplace (Last visited on August 10, 2020)

Nevada AB511: 2011: 76th Legislature, Legiscan, https://legiscan.com/NV/text/AB511/2011 (Last visited on December 26, 2020).

Nevada Passes Legislation on Autonomous Cars (Association for Unmanned Vehicle Systems International March 14, 2017), https://www.auvsi.org/nevada-passes-legislation-autonomous-cars (Last visited on October 13, 2020)

New Technologies Formation, Liability for Artificial Intelligence and Other Emerging Digital Technologies (2019), https://ec.europa.eu/transparency/regexpert/index.cfm?do=groupDetail.groupMeetingDoc&docid=36608 (Last visited on October 13, 2020)

NH user fee (toll) | ministry of road ... - morth.nic.in, (2019), https://morth.nic.in/toll?page=2 (Last visited on September 1, 2021).

Outline of systematic preparations related to autonomous driving (2018), https://www.kantei.go.jp/jp/singi/it2/kettei/pdf/20180413/auto_drive.pdf, p.6 (Last visited on April 30, 2021).

Overview of the Constitutional Challenges to the IT Act (Centre for Internet & Society), https://cis-india.org/internet-governance/blog/overview-constitutional-challenges-on-itact (Last visited on January 17, 2020)

Preliminary Report Highway HWY18MH010, https://www.ntsb.gov/investigations/AccidentReports/Reports/HWY18MH010-prelim.pdf (Last visited on December 30, 2020).

Puttaswamy v. India (Global Freedom of Expression November 24, 2014), https://globalfreedomofexpression.columbia.edu/cases/puttaswamy-v-india (Last visited on January 17, 2020)

Rachit Garg - et al., Product liability and consumer protection iPleaders (2021), https://blog.ipleaders.in/product-liability-and-consumer-protection/ (Last visited on September 1, 2021).

Rajeev Panicker, ISO 21434:2021 - an overview of newly published standard on Cybersecurity Engineering for Road Vehicles LinkedIn (2021), https://www.linkedin.com/pulse/iso-214342021-overview-newly-published-standard-road-panicker/ (Last visited on October 1, 2021).

Ramanuj, Cyber Forensics: Law and Practice in India (iPleaders October 28, 2019), https://blog.ipleaders.in/cyber-forensics-law-and-practice-in-india (Last visited on January 17, 2020)

Rayo EA, AI in Law and Legal Practice – A Comprehensive View of 35 Current Applications (Emerj November 21, 2019), https://emerj.com/ai-sector-overviews/ai-in-law-legal-practice-current-applications (Last visited on January 17, 2020)

Republican House Members Reintroduce Self Drive Act, AASHTO Journal (2020), https://aashtojournal.org/2020/09/25/republican-house-members-reintroduce-self-drive-act/ (Last visited on December 17, 2020).

Responsive Web Inc, The Fatal Accidents Act 1855 Bare acts live (1985), http://www.bareactslive.com/LCR/LC111.HTM (Last visited on September 1, 2021).

Robotics Online Marketing Team, Robotic Surgery: The Role of AI and Collaborative Robots (Robotics Online), https://www.robotics.org/blog-article.cfm/Robotic-Surgery-The-Role-of-AI-and-Collaborative-Robots/181 (Last visited on February 2, 2020)

Robotic Surgery Linked to 144 Deaths in the US (BBC News, July 22, 2015), https://www.bbc.com/news/technology-33609495 (Last visited on February 2, 2020)

Roudik P, Global Legal Monitor (Russia: Government Begins Testing Driverless Cars | Global Legal Monitor January 18, 2019), https://www.loc.gov/law/foreign-news/article/russia-government-begins-testing-driverless-cars (Last visited on October 13, 2020)

Sarrion J, Artificial Intelligence: Legal Challenges for Privacy and Data Protection (Artificial Intelligence: Legal Challenges for Privacy and Data Protection July 4, 2018), https://www.researchgate.net/publication/326331113_Artificial_Intelligence_Legal_Challenges_for_Privacy_and_Data_Protection (Last visited on January 17, 2020)

Scooter Doll & Scooter Doll Scooter Doll is a writer, Toyota suspends all self-driving vehicles at Paralympic Games after collision with athlete Electrek (2021), https://electrek.co/2021/08/27/toyota-suspends-all-self-driving-vehicles-at-paralympic-games-after-collision-with-athlete/ (last visited December 22, 2021).

Sean O'Kane, UBER DEBUTS A NEW SELF-DRIVING CAR WITH MORE FAIL-SAFES THE VERGE (2019), https://www.theverge.com/2019/6/12/18662626/uber-volvo-self-driving-car-safety-autonomous-factory-level (Last visited on December 30, 2020).

Self-Driving Car Accidents - Lawsuits (ClassAction.com October 10, 2017), https://www.classaction.com/self-driving-cars/lawsuit (Last visited on January 17, 2020)

Self-Driving Cars Get the Greenlight Under New Utah Law Government Technology State & Local Articles - e. Republic, https://www.govtech.com/policy/Self-Driving-Cars-Get-the-Greenlight-Under-New-Utah-Law.html (Last visited on December 28, 2020).

Self-driving vehicles have officially arrived in Virginia, NORTHERN VIRGINIA MAGAZINE (2019), https://northernvirginiamag.com/culture/news/2019/07/05/the-future-of-self-driving-vehicles-has-officially-arrived-in-northern-virginia/#:~:text=Perhaps%20even%20more%20importantly%2C%20a,for%20business%E2%80%9D%20for%20autonomous%20cars. (Last visited on December 28, 2020).

Self-driving vehicles Mobility, public transport and road safety | Government.nl, https://www.government.nl/topics/mobility-public-transport-and-road-safety/self-driving-vehicles#:~:text=The%20Dutch%20cabinet%20has%20adopted,driving%20vehicles)%20removes%20legal%20impediments. (Last visited on March 12, 2021).

Shinkle D and Dubois G, Autonomous Vehicles: Self-Driving Vehicles Enacted Legislation (Autonomous Vehicles | Self-Driving Vehicles Enacted Legislation), https://www.ncsl.org/research/transportation/autonomous-vehicles-self-driving-vehicles-enacted-legislation.aspx (Last visited on January 17, 2020).

Staff Writer, Back-seat driver: How Honda stole the lead in autonomous cars Nikkei Asia (2021), https://asia.nikkei.com/Spotlight/The-Big-Story/Back-seat-driver-How-Honda-stole-the-lead-in-autonomous-cars (Last visited on April 30, 2021).

State of Arkansas House Bill 1754, https://www.arkleg.state.ar.us/Acts/Document?type=pdf&act=797&ddBienniumSession=2017%2F2017R (Last visited on December 20, 2020).

State of California Senate Bill No 1298, http://www.leginfo.ca.gov/pub/11-12/bill/sen/sb_1251-1300/sb_1298_bill_20120925_chaptered.pdf (Last visited on December 22, 2020).

State of Colorado Senate Bill 17-213, http://leg.colorado.gov/sites/default/files/2017a_213_signed.pdf, http://leg.colorado.gov/sites/default/files/2017a_213_signed.pdf (Last visited on December 24, 2020).

State of Georgia House Bill No 472, https://www.legis.ga.gov/api/legislation/document/20172018/170675 (Last visited on December 24, 2020).

State of Georgia Senate Bill No 219, http://www.legis.ga.gov/Legislation/20172018/170801.pdf (Last visited on December 24, 2020).

State of Illinois HB 1341, http://iga.in.gov/legislative/2018/bills/house/1341#digest-heading (Last visited on December 24, 2020).

State of Illinois Public Act 100-0352, https://www.ilga.gov/legislation/publicacts/100/PDF/100-0352.pdf (Last visited on December 24, 2020).

State of Maine HP1204-LD1724, https://mainelegislature.org/legis/bills/getPDF.asp?paper=HP1204&item=3&snum=128 (Last visited on December 26, 2020).

State of Michigan Enrolled Senate Bill No 169, https://www.legislature.mi.gov/documents/2013-2014/publicact/pdf/2013-PA-0231.pdf (Last visited on December 26, 2020).

State of Michigan Enrolled Senate Bill No 663, https://www.legislature.mi.gov/documents/2013-2014/31 (Last visited on December 26, 2020).

State of Michigan enrolled Senate Bill No 995, https://www.legislature.mi.gov/documents/2015-2016/publicact/pdf/2016-PA-0332.pdf (Last visited on December 26, 2020).

State of Nevada Assembly Bill No 69, https://www.leg.state.nv.us/Session/79th2017/Bills/AB/AB69_EN.pdf (Last visited on December 26, 2020).

State of Nevada Senator Bill No 313, https://www.leg.state.nv.us/Session/77th2013/Bills/SB/SB313_EN.pdf (Last visited on December 26, 2020).

State of New York S 2005-C, A 2005-C, https://nyassembly.gov/2017budget/budget_bills/A3005C.pdf (Last visited on December 26, 2020).

State of Tennessee Senate Bill No 1561, https://publications.tnsosfiles.com/acts/109/pub/pc0927.pdf (Last visited on December 28, 2020).

State of Texas HB 1791, https://capitol.texas.gov/tlodocs/85R/billtext/pdf/HB01791F.pdf#navpanes=0 (Last visited on December 28, 2020).

State of Texas SB 2205, https://capitol.texas.gov/tlodocs/85R/billtext/pdf/SB02205F.pdf#navpanes=0 (Last visited on December 28, 2020).

Strategie automatisiertes und vernetztes Fahren BMVI, https://www.bmvi.de/SharedDocs/DE/Publikationen/DG/broschuere-strategie-automatisiertes-vernetztes-fahren.html (Last visited on March 2, 2021)

Team GreyB, Top 30 Autonomous Vehicle Technology and Car Companies GreyB (2021), https://www.greyb.com/autonomous-vehicle-companies/ (Last visited on Jun 25, 2021).

Tennessee SB151, TRACK BILL, https://trackbill.com/bill/tennessee-senate-bill-151-motor-vehicles-as-enacted-enacts-the-automated-vehicles-act-and-other-requirements-related-to-the-operation-of-autonomous-vehicles-on-the-public-roads-of-this-state-amends-tca-title-5-title-6-title-7-title-39-title-40-title-54-title-55-title-56-title-65-and-title-67/1375094/#/details=true (Last visited on December 28, 2020).

Tesla Auto Pilot Lawsuit, https://media.forthepeople.com/wp-content/uploads/2018/10/Tesla-autopilot-lawsuit-complaint.pdf (Last visited on January 17, 2020)

Tesla Autopilot, WIKIPEDIA (2021), https://en.wikipedia.org/wiki/Tesla_Autopilot#Handan,_China_(January_20,_2016) (Last visited on January 2, 2021).

Tesla fatal autopilot crash: family may have grounds to sue, legal experts say, THE GUARDIAN (2016), https://www.theguardian.com/technology/2016/jul/06/tesla-autopilot-crash-joshua-brown-family-potential-lawsuit (Last visited on January 2, 2021).

Text - H.R.8350 - 116th congress (2019-2020): Self Drive, https://www.congress.gov/bill/116th-congress/house-bill/8350/text (last visited December 22, 2021).

The Jakarta Post, Jokowi wants all vehicles in new capital city to be autonomous, electric The Jakarta Post, https://www.thejakartapost.com/news/2020/01/15/jokowi-wants-all-vehicles-in-new-capital-city-to-be-autonomous-electric.html (Last visited on April 30, 2021).

The rise of autonomous vehicles | Digital Watch (2020), https://dig.watch/trends/rise-autonomous-vehicles (Last visited on March 2, 2021).

The RMLNLU Law Review Blog, Moving towards a driverless era The RMLNLU Law Review Blog (2021), https://rmlnlulawreview.com/2017/10/25/moving-towards-a-driverless-era/ (Last visited on October 1, 2021).

The state of autonomous legislation in Europe | Autovista Group, https://autovistagroup.com/news-and-insights/state-autonomous-legislation-europe (Last visited on March 22, 2021)

Towards a European Regulation of Autonomous Vehicles - EU Perspectives and the German Model, https://www.researchgate.net/publication/333575970_Towards_a_European_Regulation_of_Autonomous_Vehicles_-_EU_Perspectives_and_the_German_Model (Last visited on March 2, 2021).

Tuk-tuk in Thailand to get a makeover, government plans to transform commute with autonomous vehicles, The Economic Times, https://economictimes.indiatimes.com/magazines/panache/tuk-tuk-in-thailand-to-get-a-makeover-government-plans-to-transform-commute-with-autonomous-vehicles/articleshow/71389064.cms?from=mdr (Last visited on May 5, 2021).

Uber gets permit to restart testing its self-driving cars in California, REUTERS (2020), https://www.reuters.com/article/us-uber-self-driving/uber-gets-permit-to-restart-testing-its-self-driving-cars-in-california-idUSKBN1ZZ2QG (Last visited on December 30, 2020).

USDOT Comprehensive Management Plan for Automated Vehicle Initiatives, U.S. Department of Transportation, https://www.transportation.gov/policy-initiatives/automated-vehicles/usdot-comprehensive-management-plan-automated-vehicle (Last visited on December 17, 2020).

Utah HB0101: 2019: General Session, LEGISCAN, https://legiscan.com/UT/text/HB0101/2019 (Last visited on December 28, 2020).

Vermont Laws, State House Dome, https://legislature.vermont.gov/statutes/fullchapter/23/041 (Last visited on December 28, 2020).

Việt Nam's first autonomous vehicle debuts, vietnamnews.vn, https://vietnamnews.vn/economy/914897/viet-nams-first-autonomous-vehicle-debuts.html (Last visited on May 10, 2021).

What is an autonomous car? – how self-driving cars work, Synopsys, https://www.synopsys.com/automotive/what-is-autonomous-car.html (last visited December 12, 2021).

What is Truck Platooning (2017), https://www.acea.be/uploads/publications/Platooning_roadmap.pdf (Last visited on December 22, 2020).

William Thornton | wthornton@al.com, Alabama Legislature preparing for self-driving cars AL (2019), https://www.al.com/news/anniston-gadsden/2019/03/how-the-alabama-legislature-is-preparing-for-self-driving-cars.html (Last visited on December 17, 2020).

www.ETAuto.com, Autonomous cars to grow to 24 million units in Asia-Pacific by 2024 - ET Auto ETAuto.com (2019), https://auto.economictimes.indiatimes.com/news/auto-technology/autonomous-cars-to-growth-to-24-million-units-in-asia-pacific-by-2024/70701274 (Last visited on April 25, 2021).

f. Newspaper Articles

Grady D, A.I. Comes to the Operating Room (The New York Times January 6, 2020), https://www.nytimes.com/2020/01/06/health/artificial-intelligence-brain-cancer.html (Last visited on January 17, 2020).

Lu D, DeepMind Found an AI Learning Technique Also Works in Human Brains" (New Scientist January 15, 2020), https://www.newscientist.com/article/2230327-deepmind-found-an-ai-learning-technique-also-works-in-human-brains (Last visited on January 17, 2020).

Opinion: Applying AI to Clinical Care Is Key to Individualized Medicine (The Scientist Magazine®), https://www.the-scientist.com/reading-frames/opinion—applying-ai-to-clinical-care-is-key-to-individualized-medicine-66126 (Last visited on January 17, 2020).

Zargar H, India's Information Technology Act Has Not Been Effective in Checking Cyber Crime: Expert (DNA India April 3, 2013), https://www.dnaindia.com/technology/report-india-s-information-technology-act-has-not-been-effective-in-checking-cyber-crime-expert-1818328 (Last visited on January 20, 2020).